Profiles in Flue Gas Desulfurization

Dr. Richard R. Lunt, Arthur D. Little, Inc.; and
John D. Cunic, ExxonMobil Research and Engineering Company

AIChE
American Institute of
Chemical Engineers
www.aiche.org

Published by the American Institute of Chemical Engineers
3 Park Avenue, New York, NY 10016-5991 USA
in cooperation with its
Center for Waste Reduction Technologies (CWRT), an
Industry Technology Alliance

© 2000 by the American Institute of Chemical Engineers and its
Center for Waste Reduction Technologies
3 Park Avenue, New York, NY 10016-5991

Library of Congress Catalog Card Number 00-036222
ISBN: 0-8169-0820-6

Coordinating Editor: Beth Shery Sisk
Cover design and typesetting: Terry A. Baulch

Lunt, Richard R., 1944-
Profiles in flue gas desulfurization/by Richard R. Lunt and John D. Cunic
p. cm.
ISBN 0-8169-0820-6
1. Flue gases--Desulfurization. I. Cunic, John D., 1946- II. Title.

TD885.5.S8 L85 2000
628.5'32--dc21   00-036229

All rights reserved. No part of this publication may be reproduced, stored in a retrieval system, or transmitted in any form or by any means, electronic, mechanical, photocopying, recording, or otherwise, without the prior permission of the American Institute of Chemical Engineers.

SECTION 1

Profiles in Flue Gas Desulfurization

Purpose

This review of FGD (Flue Gas Desulfurization) provides tabular summaries of technology as of early 1999 and a compendium of process profiles. Each profile provides a characterization of the technology with respect to basic equipment components, performance capabilities, operational requirements, range of applications and potential limitations.

Scope and Content

Technologies Included

Technologies included in this compendium are for "tail-end" SO_2 control processes that have either achieved commercial application or an advanced stage of development. Hence, all included technologies fulfill one of the following two criteria:

1) Commercialized – The technology has been applied to a commercial operation at a scale of about 250,000 scfm (100 MW equivalent) or greater. Discretion is used taking into account the number, size, age and success of applications.

2) Developmental Stage – The technology has successfully progressed through a full pilot scale test program, usually 5,000-10,000 scfm (2 to 4 MW equivalent) and, preferably, through demonstration testing, typically 50,000-100,000 scfm (20 to 40 MW equivalent).

Process technologies are summarized in Tables 1 and 2. These are divided into two groups – Waste Producing Processes and Byproduct Processes. In Waste Producing Processes, absorbed SO_2 is discharged either as a neutralized solution of a soluble salt or disposed as a dewatered solid precipitate (usually calcium sulfite or sulfate), neither of which typically has significant economic value or reuse potential. Byproduct processes convert absorbed SO_2 to a form having reuse potential or market value (either directly or after further processing). Such byproduct forms include commercial grade gypsum, fertilizer, sulfuric acid, elemental sulfur, and SO_2-rich gas that can be converted to acid or sulfur. Within each of these major groups, processes are ordered first according to those that are Commercial, and then those that are Developmental. Profiles are presented in the same general order.

Technologies Excluded

Process technologies excluded from this compendium fall into one of the three following categories.

1) Conceptual Stage – The technology is in a very early stage of development. It has not pro-

gressed through pilot scale testing with significant success.

2) Abandoned Development – Technology development has been suspended or abandoned for more than 15 years and/or has demonstrated significant performance shortfalls or economic barriers.

3) Not traditional FGD – The technology does not involve traditional gas phase SO_2 control. Examples in this category include: fluid bed boilers, H_2S control (hot gas desulfurization for advanced power systems), and SO_2 control integral to chemical and gas processing such as certain amine or solvent technologies used exclusively in refining applications.

Table 3 summarizes process technologies excluded either because they have not moved beyond early development or development was abandoned more than 15 years ago.

Profiles

Presentation

Format: Each technology profile consists of two pages — a one-page process schematic depicting the major equipment components and arrangement, and a one-page narrative outlining its features and applications. Where a technology has more than one process offering (say, from more than one supplier) they are grouped together. As space permits, significant variations are noted.

Groupings: There are several technologies that fall under both the Waste Producing and Byproduct categories, with the specific application and mode of operation determining which is applicable. In such cases, they are shown in both listings of Tables 1 and 2; however, only one profile is provided. Notable examples include: Limestone Slurry Forced Oxidation, Limestone Clear Solution Scrubbing, and Sodium/Lime Dual Alkali (Dilute Mode). There are also some technology categories where a few processes have been commercialized, but others remain developmental. Where at least one process has been commercialized and remains active, the entire technology category is considered Commercial.

Integrated NO_x Control: An intentional differentiation has been made between those technologies capable of SO_2 control only and those capable of "integrated" SO_2 and NO_x control. In this context, "integrated" means that NO_x control is incorporated within the overall process system and affects one or more of the following: system configuration; equipment requirements; mode of operation; process chemistry; or performance characteristics. An obvious example would be where both SO_2 and NO_x control are accomplished within a single vessel. Simply "coupling" a selective catalytic reduction (SCR) unit for NO_x control on the front end of an FGD process is not considered integration.

Conventions and Clarifications

Process Characterizations

Other Gaseous Emissions: Point source emissions of gaseous compounds (not particulate) deemed to be of potential significance are noted. This includes both separate process unit vents as well as impacts to the main stack emissions as a result FGD operations. Only emissions associated with the battery limits of the FGD process itself are noted. Emissions from "downstream" conversion processes for converting SO_2-rich gas to sulfuric acid or elemental sulfur are not.

Inlet SO_2 Levels: For commercial technologies, the range is for commercial operations to the extent known. Frequently, testing of processes extends beyond the range of commercial operations. Where testing is known to have demonstrated a broader range of capability, this is listed.

NO$_x$ Removal: Capability for NO$_x$ control as an integral part of the process is indicated (the definition of "integral" is provided above). Where development of integrated control is active and in advanced stages, this is indicated.

Particulate Removal: There are two indications regarding particulate control. First, is the capability for particulate control as an integral part of the process. In this context, "integral" means that the control is incorporated within the process itself rather than as a separate gas treatment system coupled on one end or the other. An obvious example would be the use of a high-efficiency wet scrubber for combined particulate and SO$_2$ control in a waste producing process. Second, is the notation where there is a need for decoupled, high-efficiency particulate control upstream of the SO$_2$ absorber.

Commercial Applications

A Commercial technology is considered "Active" if at least one commercial installation is still in operation. A Developmental technology is considered "Active" if R&D efforts are continuing or have been active within the past three years and opportunities for advancement are being pursued.

Process Description

Process descriptions are intended to be rather general. Where there are a number of processes within a technology category, descriptions generally cover the most common forms of the processes. Where there are significant variations, they will usually be mentioned, space permitting.

Advantages/Disadvantages

Advantages, disadvantages and limitations are the authors' understanding and knowledge and are generally attributed to the technology as a whole for typical applications. Where there are many processes within a technology category, some processes may not exhibit all of these. Each specific application must be evaluated in terms of "degree of fit" for any technology or individual process.

Principal Suppliers/Developers

Readers are cautioned that lists of process suppliers or developers reflect representations as of early 1999. They are not intended to be exhaustive and, due to the fluid nature of this business, are subject to change. Omission of a supplier is not intended to imply that the supplier is incapable of providing the technology; nor does inclusion guarantee that a supplier still can or will supply a given process. Where process names are known or are significant in identifying processes, these are provided.

Table 1A. FGD Technology Summary Matrix - Waste Producing Processes

Technology Descriptor	Name(s) of Principal Process(es)	Page Number	Status	FGD Only	Integrated FGD & DeNOx
Commercialized Technology					
Solid Waste Processes					
Conventional Lime Slurry	Generic	10-11	Active	●	○
Lime Slurry - Magnesium-Promoted	Generic	12-13	Active	●	○
Lime Slurry Forced Oxidation	Generic	52-53	Active	●	
Limestone Slurry Natural Oxidation	Generic	14-15	Active	●	○
Limestone Slurry Inhibited Oxidation	Generic	16-17	Active	●	○
Limestone Slurry Forced Oxidation	Many/Generic	54-55	Active	●	
Limestone Slurry - Magnesium-Promoted	Generic	18-19	Active	●	○
Alkaline Ash Scrubbing	Generic	20-21	Active	●	
Sodium/Lime Dual Alkali (Concentrated Mode)	Generic	22-23	Active	●	
Sodium/Lime Dual Alkali (Dilute Mode)	Generic	58-59	Active	●	
Spray Dry Absorption - Lime-Based	Many/Generic	24-25	Active	●	
Spray Dry Absorption - Sodium-Based	Generic	26-27	Active	●	
Circulating Fluid/Entrained Bed - Lime-Based	CDS; GSA; et al	28-29	Active	●	
Sodium-Based Duct Injection	Generic	30-31	Active	●	
Furnace Sorbent Injection	ARA; DISCUS; LIFAC; LIMB R-SOX; SONOX; TAV; et. al.	32-33	Active	●	○
Liquid Waste Processes					
Sodium Solution - Once-Through	Generic	34-35	Active	●	○
Magnesium Hydroxide - Once-Through	Generic	36-37	Active	●	
Seawater Scrubbing	Generic	38-39	Active	●	

● Principal focus of all processes within technology category
○ Some processes have integrated DeNO$_x$ capability or additives are in advanced state of development for some processes

Active - Commercial systems are still in operation or development is continuing
Inactive - No commercial systems are known to still be in operation or development has been suspended

Table 1B. FGD Technology Summary Matrix - Waste Producing Processes

Technology Descriptor	Name(s) of Principal Process(es)	Page Number	Status	FGD Only	Integrated FGD & DeNOx
Developmental Technology					
Solid Waste Processes					
Limestone Clear Liquor Scrubbing	Generic	96-97	Active	●	○
Sodium/Limestone Dual Alkali (Concentrated Mode)	Generic	40-41	Inactive	●	
Circulating Fluid/Entrained Bed with NO$_x$ Control	CDS	42-43	Active		●
Lime-Based Duct Injection	ADVACATE; Coolside; CZD; E-SOX; HALT; HYPAS; LILAC	44-45	Active	●	
Lime-Based Economizer Injection	SO$_x$-NO$_x$-Rox-Box; EI	46-47	Active	●	○
Liquid Waste Processes					
Condensing Heat Exchanger	IFGT	48-49	Active	●	

● Principal focus of all processes within technology category
○ Some processes have integrated DeNO$_x$ capability or additives are in advanced state of development for some processes

Active - Commercial systems are still in operation or development is continuing
Inactive - No commercial systems are known to still be in operation or development has been suspended

Table 2A. FGD Technology Summary Matrix - Byproduct Processes

Technology Descriptor	Name(s) of Principal Process(es)	Page Number	Status	FGD Only	Integrated FGD & DeNOx
Commercialized Technology					
Gypsum Processes					
Lime Slurry Forced Oxidation	Generic	52-53	Active	●	
Limestone Slurry Forced Oxidation	Many/Generic	54-55	Active	●	
Dilute Sulfuric Acid to Gypsum	Chiyoda 101	56-57	Inactive	●	
Sodium/Lime Dual Alkali (Dilute Mode)	Generic	58-59	Active	●	
Sodium/Limestone Dual Alkali with H_2SO_4 Conversion	Generic	60-61	Active	●	
Aluminum Sulfate/Limestone Dual Alkali	Dowa	62-63	Active	●	
Ammonia/Lime Dual Alkali	Kurabo	64-65	Active	●	
Magnesium Solution/Lime Dual Alkali	Thioclear	66-67	Active	●	
Magnesium Slurry/Lime Dual Alkali	Kawasaki	68-69	Active	●	
Fertilizer Processes					
Ammonia Scrubbing - Once-Through	ATS	70-71	Active	●	
Ammonia Scrubbing with Oxidation	Generic	72-73	Active	●	
Ammonia Scrubbing with SCR	Walther	74-75	Active		●
Electron Beam Irradiation	E-Beam; Pulse Energization	76-77	Active		●
Sulfuric Acid Processes					
Ammonia Scrubbing with Acid Regeneration	Cominco	78-79	Active	●	
Magnesium Slurry with Thermal Regeneration	Magnesium Oxide Recovery	80-81	Active	●	
Direct Sulfuric Acid Conversion	Cat-Ox; WSA	82-83	Active	●	
Direct Sulfuric Acid Conversion with NO_x Control	DESONOX; SNOX	84-84	Active		●
SO_2-Rich Gas Processes					
Cold Water Scrubbing with Thermal Sripping	Boliden Process	86-87	Active	●	
Sodium Sulfite with Thermal Regeneration	Wellman Lord	88-89	Active	●	
Organic Solvent with Thermal Regeneration	Solinox	90-91	Active	●	
Amine Solution with Thermal Regeneration	DMA; Sulphidine	92-93	Inactive	●	
Activated Carbon with Thermal Regeneration	GE-Mitsui-BF; EPDC/SHI	94-95	Active		●

● Principal focus of all processes within technology category
○ Some processes have integrated $DeNO_x$ capability or additives are in advanced state of development for some processes

Active - Commercial systems are still in operation or development is continuing
Inactive - No commercial systems are known to still be in operation or development has been suspended

Table 2B. FGD Technology Summary Matrix - Byproduct Processes

Technology Descriptor	Name(s) of Principal Process(es)	Page Number	Status	FGD Only	Integrated FGD & DeNOx
Developmental Technology					
Gypsum Processes					
Limestone Clear Liquor Scrubbing	Generic	96-97	Active	●	
Sodium Acetate/Limestone Dual Alkali	Kureha	98-99	Inactive	●	
Fertilizer Processes					
Sodium (Bi)Carbonate Sorption with Ammonia Regeneration	Airborne Technologies	100-101	Active	●	●
Ammonium & Calcium Pyrophosphate Scrubbing	Pircon Peck	102-103	Inactive	●	
Scrubbing with Cement Kiln Waste Dust	Passamaquoddy Recovery Process	104-105	Inactive	●	
Sulfuric Acid Processes					
Bromine Solution with Electrolytic Regeneration	ISPRA	106-107	Unknown	●	
Electrochemical Membrane Separation	Generic	108-109	Active	●	
Sulfuric Acid Absorption with Peroxide Oxidation	Peroxide	110-111	Unknown	●	
Carbon Adsorption with Acid Regeneration	Generic	112-113	Inactive	●	
Zinc Oxide Slurry with Thermal Regeneration	ZnO (Batelle)	114-115	Active		●
SO₂-Rich Gas Processes					
Sodium Phosphate with Thermal Regeneration	ELSORB	116-117	Inactive	●	
Amine Solution with Thermal Regeneration	CANSOLV; Dow; NOSOX	92-93	Inactive	●	○
Ammonia with Thermal Regeneration	Stackpol 150; Exorption; ABS	118-119	Inactive	●	
Sodium Sulfite with Solvent Extraction	Tung	120-121	Inactive	●	
Sodium Hydroxide with Electrolytic Regeneration	Ionics	122-123	Inactive	●	
Sodium Sulfite with Electrodialysis Regeneration	SOXAL	124-125	Inactive		●
Sodium Sulfite with Zinc Oxide & Thermal Regeneration	ZnO (Univ. of Illinois)	126-127	Inactive	●	
Magnesia/Vermiculite Adsorption	Sorbtech	128-129	Active		●
Sulfur Processes					
Sodium Citrate with Liquid Claus Regeneration	Citrate	130-131	Inactive	●	
Sodium Phosphate with Thermal Regeneration	Sulf-X	132-133	Inactive	●	○
Direct Gas Phase Reduction	Parsons	134-135	Inactive		●
Copper Oxide Sorption with Reduction Regeneration	COBRA; CuO	136-137	Active		●●●
Alkalized Alumina Adsorption	NOXSO	138-139	Inactive		●

● Principal focus of all processes within technology category
○ Some processes have integrated DeNOx capability or additives are in advanced state of development for some processes

Active - Commercial systems are still in operation or development is continuing
Inactive - No commercial systems are known to still be in operation or development has been suspended

Table 3. FGD Technology Summary Matrix - Processes Not Included

Technology Descriptor	Name(s) of Principal Process(es)	Status	Exclusion Commentary
Solid Waste Processes			
Dry Limestone Moving Bed	LEC	Developmental	Very early development stage
Limestone Scrubbing with Phosphorus Additives	Phos/NOx	Developmental	Very early development stage
Byproduct Processes			
Soda Ash Dry Injection with Sorbent Regeneration	ENELCO	Developmental	Very early concept development - not piloted
Sodium Carbonate with Coke Regeneration	Molten Carbonate	Developmental	Demonstration failed; abandoned in early 1980s
Sodium Phosphate with Aqueous Claus	Stauffer	Developmental	Development abandoned in early 1980s
Biological Reduction	BIO-FGD	Developmental	Very preliminary conceptual development
Gas Phase Thermal Oxidation	Plasma-Jet Treatment	Developmental	Very early development stage
Sodium Sulfide Scrubbing with Barium Regeneration	SULFRED	Developmental	Development abandoned in mid 1980s
Carbon Adsorption + Reduction with Coke to Sulfur	FW-BF/RESOX; Westvaco	Developmental	Development abandoned in mid 1980s

SECTION 2

FGD TECHNOLOGY PROFILES
Waste Producing Processes

Conventional Lime Slurry

Commercial
Absorption of SO₂ in lime slurry followed by solids separation, dewatering and often a mixture of the solids with fly ash and lime for disposal.

Process Characteristics
SO₂ Sorbent: A mixture of lime and calcium sulfite/sulfate
Principal Raw Materials: Lime; Additives in some applications (e.g., thiosulfate, emulsified sulfur, formate)
Potentially Saleable Byproducts: None
Solid Wastes: Mixed calcium sulfite and sulfate waste solids sometimes admixed with fly ash (and lime)
Liquid Effluents: None
Other Gaseous Emissions: None
Inlet SO$_2$ Levels: <100 to 6,500 ppm (Commercial Operations)
SO$_2$ Removal: Typical Design Range: 90-95%; Maximum at Highest Inlet SO$_2$: ~98%
NO$_X$ Removal: Nil - Although additives are under development that can be used for integrated control (e.g., soluble iron chelates)
Particulate Removal: Integrated capability, but need to mix dry ash with FGD waste in some applications limits coupling of particulate control

Commercial Application
Number/Types: Many: Utility & industrial boilers; chemical process units; metallurgical smelters and furnaces; general manufacturing units
Locations: Global
First Deployment: 1970s (Current generation of technology)
Current Status: Active

Process Description
In a typical system, the gas contactor consists of a multi-level spray tower or deluge type device (e.g., venturi scrubber) or combination of both with an internal recirculation tank at the bottom. Lime is fed to the recirculation tank and slurry from the tank is recirculated through the contactor at a high L/G ratio. Slurry is continuously withdrawn from the scrubber and dewatered by conventional thickening and filtration. In many systems on coal-fired boilers, the dewatered solids are admixed with fly ash and lime to improve handling and promote solidification for disposal. An option is to pump thickened solids directly to onsite ponds where the solids are settled out and the clarified liquor returned to the system. This practice, though, is now primarily employed only in systems located at more remote sites such as metallurgical plants. In some applications, additives such as thiosulfate or emulsified sulfur are added to the absorbent solution to reduce oxidation in order to prevent calcium sulfate scaling problems.

Advantages/Disadvantages

Advantages
- Well developed technology with many suppliers

Disadvantages/Limitations
- Produces a solid waste of little reuse potential
- Solid waste usually requires admixture with dry ash for landfill disposal

Principal Suppliers

ABB	B&W	Nippon Kokan	Saarberg-Holter	UOP
AirPol/FLS	Deutsche Babcock	Procedair	Thyssen	Wheelabrator
American Air Filter	Marsulex	Research Cottrell		
Anderson 2000	Mitsubishi	Riley		

Waste Producing Processes

Conventional Lime Slurry / Commercial

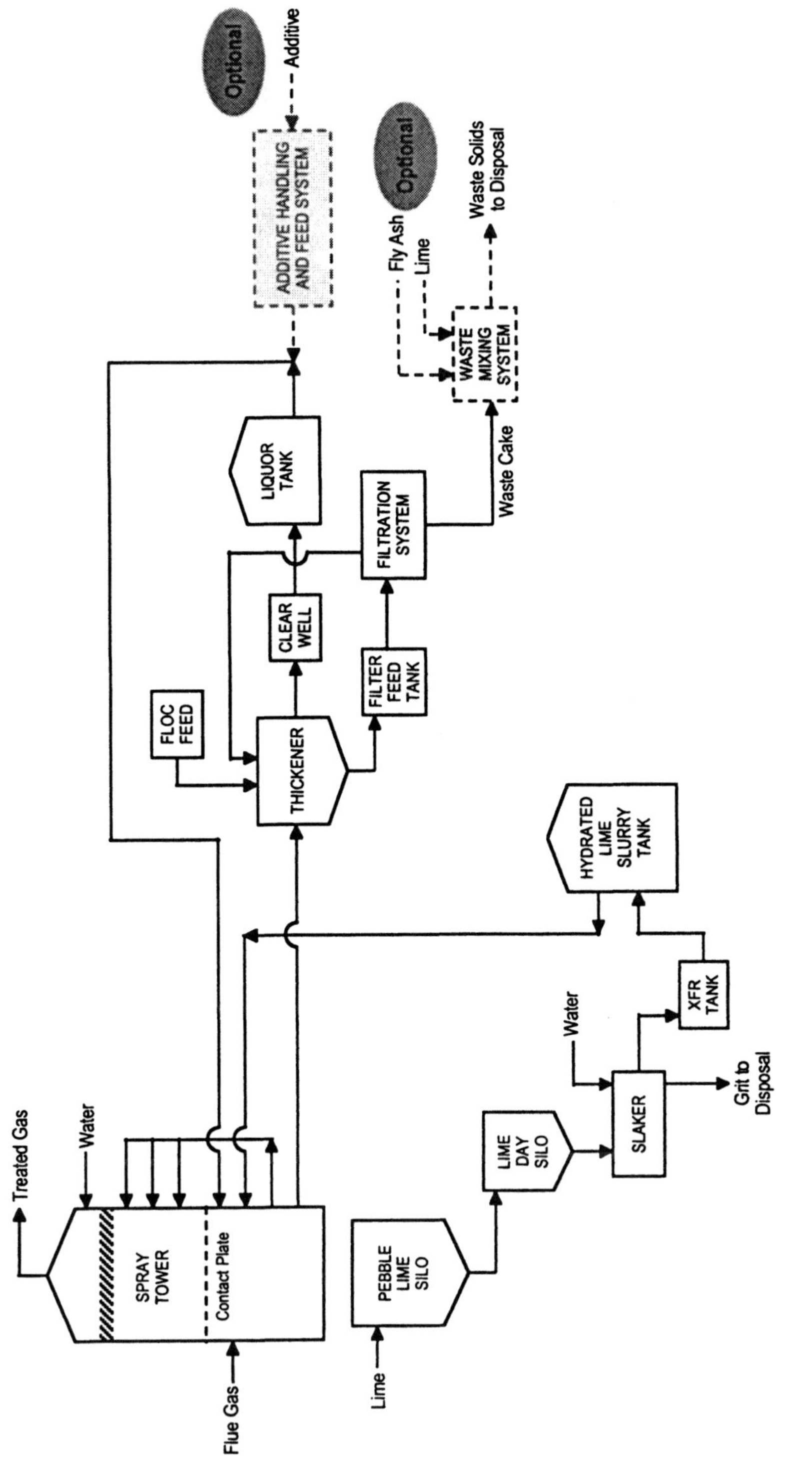

Lime Slurry – Magnesium-Promoted

Commercial
Absorption of SO$_2$ in lime slurry containing dissolved magnesium salts, followed by solids separation and mixture of waste solids with ash and lime for disposal.

Process Characteristics
SO$_2$ Sorbent: A mixture of lime and dissolved magnesium salts
Principal Raw Materials: Lime (usually high magnesium content such as Dravo's Thiosorbic, lime; Magnesium oxide or hydroxide if calcitic lime is used
Potentially Saleable Byproducts: None
Solid Wastes: A mixed calcium sulfite and sulfate waste solids usually admixed with fly ash (and lime)
Liquid Effluents: None
Other Gaseous Emissions: None
Inlet SO$_2$ Levels: ~1,000 to 4,000 ppm (Commercial Operations)
SO$_2$ Removal: Typical Design Range: 90-95%; Maximum at Highest Inlet SO$_2$: ~99%
NO$_X$ Removal: Nil - Although additives are under development that can be used for integrated control (e.g., soluble iron chelates)
Particulate Removal: Integrated capability, but need to mix dry ash with FGD waste in some applications limits coupling of particulate control

Commercial Application
Number/Types: Many: Mostly utility & industrial boilers
Locations: US; Asia
First Deployment: Late 1970s
Current Status: Active

Process Description
In a typical system, the gas contactor consists of a multi-level spray tower or deluge scrubber (e.g., venturi) or combination of both with an internal recirculation tank at the bottom. Slurry of lime and magnesium hydroxide is fed to the recirculation tank which is recirculated through the contactor at a high L/G ratio. The presence of magnesium increases the alkalinity of the solution enhancing SO$_2$ removal. Slurry is continuously withdrawn from the scrubber and dewatered usually by conventional thickening and filtration. In most systems on coal-fired boilers, the dewatered solids are admixed with fly ash and lime to improve handling and promote solidification for disposal because the solids typically cannot be dewatered as well as those produced in systems without magnesium addition. An option is to pump thickened solids directly to onsite ponds where solids are settled out and clarified liquor returned to the system, although this practice has been generally phased out. There are three approaches to providing the magnesium hydroxide used in the absorption. Dravo, which originally developed and commercialized the technology, sells a high magnesium content calcitic lime (4-7% MgO). Bechtel has developed technology based upon the use of dolomitic lime (3 -6 % MgO) in combination with calcitic lime, although this requires the use of two different slaking systems. Finally, Mg(OH)$_2$ can be separately added either as purchased or from MgO slaked onsite.

Advantages/Disadvantages

Advantages
- Use of magnesium alkali addition increases the liquor alkalinity which enhances SO$_2$ removal and reduces sulfate formation potential

Disadvantages/Limitations
- Produces a solid waste of little reuse potential
- Solid waste usually requires admixture with dry ash for landfill disposal

Principal Suppliers
| B&W | Bechtel | Dravo | Marsulex |

Waste Producing Processes

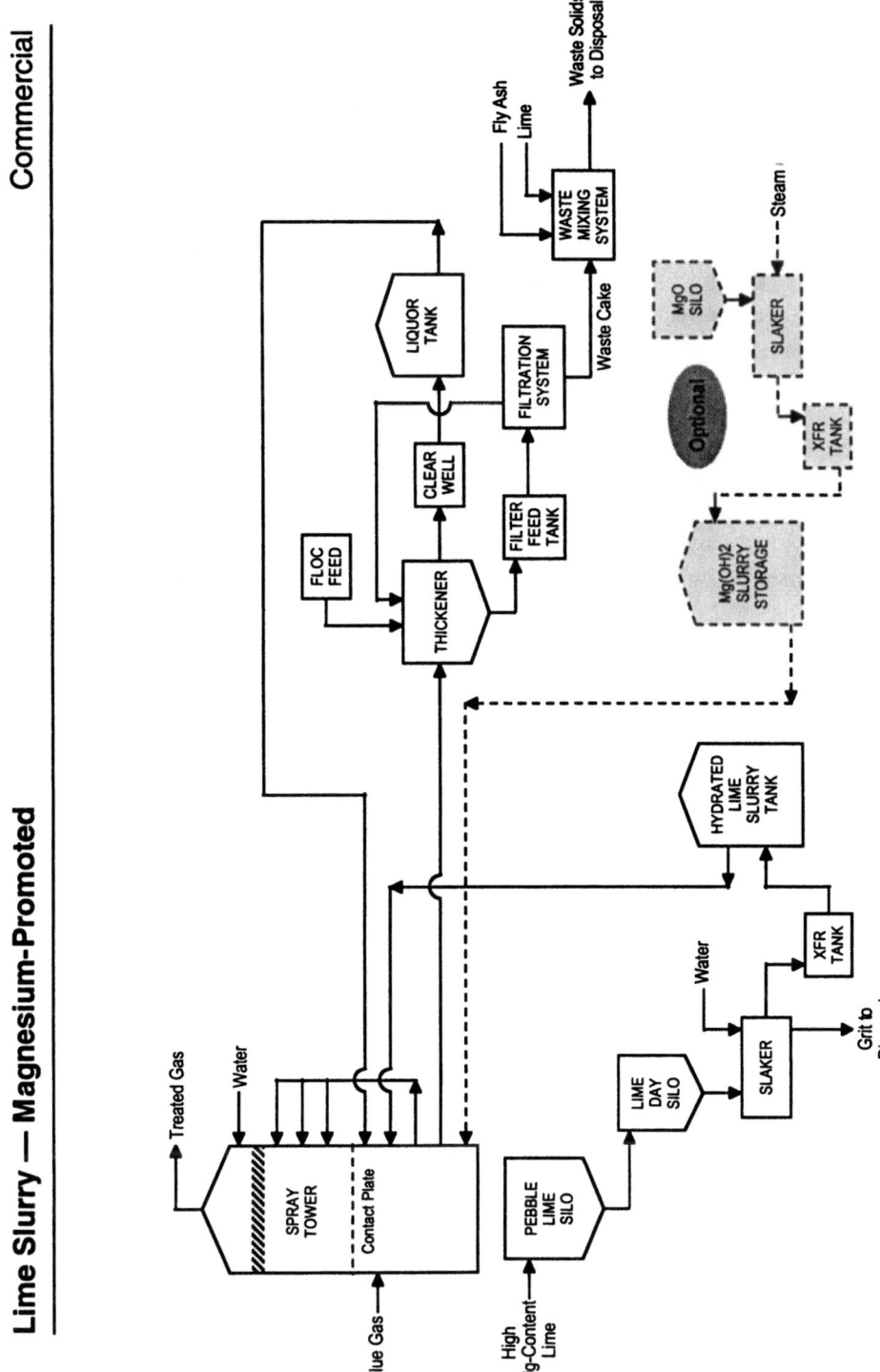

Lime Slurry — Magnesium-Promoted Commercial

Limestone Slurry Natural Oxidation

Commercial
Absorption of SO₂ in a limestone slurry precipitating mixed calcium sulfite/sulfate solids that are continuously withdrawn and dewatered either in onsite ponds or by conventional thickening and filtration for landfill.

Process Characteristics
SO$_2$ Sorbent: A mixture of limestone, calcium sulfite/sulfate solids
Principal Raw Materials: Limestone
Potentially Saleable Byproducts: None
Solid Wastes: A mixed calcium sulfite/sulfate waste solids
Liquid Effluents: None
Other Gaseous Emissions: None
Inlet SO$_2$ Levels: <1,500 ppm (Commercial Operations)
SO$_2$ Removal: Typical Design Range: 75-90%; Maximum at Highest Inlet SO$_2$: ~80-85%
NO$_X$ Removal: Nil - Although additives are under development that can be used for integrated control (e.g., soluble iron chelates)
Particulate Removal: Integrated control capability

Commercial Application
Number/Types: Many: Mostly utility & industrial boilers
Locations: North America; Europe
First Deployment: 1960s (Current generation of technology)
Current Status: Active

Process Description
In a typical system, the gas contactor consists of a multi-level spray tower or a combination deluge scrubber/spray tower with an internal recirculation tank at the bottom. Limestone is fed to the recirculation tank and slurry from the tank is recirculated through the sprays at a high L/G ratio. Slurry is continuously withdrawn from the scrubber and dewatered with recovered liquor returned to the scrubber and for use in limestone slurry preparation. Most early systems (1970s) utilized onsite ponds for impounding waste solids with clarified liquor returned to the system for scrubber makeup and preparation of slurried limestone. Later systems (late 1970s-early 1980s) utilized conventional dewatering equipment, usually thickener/clarifiers and vacuum filters. Systems installed on applications greater than 1,000 ppm SO$_2$ were plagued with either excessive scaling or low SO$_2$ removal or both, and were upgrade/replaced with either inhibited oxidation or forced oxidation techniques discussed later.

Advantages/Disadvantages

Advantages
- Uses low cost limestone reagent

Disadvantages/Limitations
- Generally limited to <~1,000 ppm inlet SO$_2$
- Removal efficiencies limited to ~85% at highest inlet SO$_2$ levels
- Produces a solid waste of little reuse potential

Principal Suppliers
ABB	Deutsche Babcock	Research Cottrell
B&W	Marsulex	Riley

Waste Producing Processes

Limestone Slurry Natural Oxidation — Commercial

NOTE: Direct ponding of either scrubber blowdown or thickened solids is commonly practiced.

Limestone Slurry Inhibited Oxidation

Commercial
Absorption of SO$_2$ in limestone slurry buffered with organic acid, followed by solids separation and usually admixture of the solids with fly ash and lime for disposal.

Process Characteristics
SO$_2$ Sorbent: A mixture of limestone, calcium sulfite and organic buffering agent
Principal Raw Materials: Limestone; Organic acid (dibasic acid)
Potentially Saleable Byproducts: None
Solid Wastes: A mixed calcium sulfite and sulfate waste solids usually admixed with fly ash (and lime)
Liquid Effluents: None
Other Gaseous Emissions: None
Inlet SO$_2$ Levels: ~500 to ~3,500 ppm (Commercial Operations)
SO$_2$ Removal: Typical Design Range: 90-95%; Maximum at Highest Inlet SO$_2$: ~98%
NO$_x$ Removal: Nil - Although additives are under development that can be used for integrated control (e.g., soluble iron chelates)
Particulate Removal: Integrated control capability, but admixture of ash with FGD waste is usually required which limits coupling of particulate control

Commercial Application
Number/Types: Many: Mostly utility & industrial boilers
Locations: North America; Europe
First Deployment: 1980s (Current generation of technology)
Current Status: Active

Process Description
In a typical system, the gas contactor consists of a multi-level spray tower with an internal recirculation tank at the bottom. Limestone is fed to the recirculation tank and slurry from the tank is recirculated through the sprays at a high L/G ratio. Slurry is continuously withdrawn from the scrubber and dewatered by conventional thickening and filtration. In many systems on coal-fired boilers, the dewatered solids are admixed with fly ash and lime to improve handling and promote solidification for disposal. Additives, including organic dibasic acids or formic acid and emulsified sulfur, are added to to make up for losses to the waste and degradation within the system. Organic acids buffer the solution enhancing SO$_2$ removal capacity; and sulfur reduces oxidation which prevents calcium sulfate scaling problems.

Advantages/Disadvantages

Advantages
- Uses low cost limestone reagent
- Use of buffering agents and emulsified sulfur control the system chemistry minimizing scaling problems and enhancing SO$_2$ removal efficiency

Disadvantages/Limitations
- Generally limited to ~4,000 ppm inlet SO$_2$ although designs are proposed for higher levels
- Removal efficiencies limited to ~97% at highest inlet SO$_2$ levels
- Produces a solid waste of little reuse potential
- Solid waste typically requires admixture with dry ash for disposal in landfills

Principal Suppliers

ABB	Lentjes	Research Cottrell	Wheelabrator
B&W	Marsulex	Riley	
Deutsche Babcock	Nippon Kokan	UOP	

Waste Producing Processes

Limestone Slurry Inhibited Oxidation — Commercial

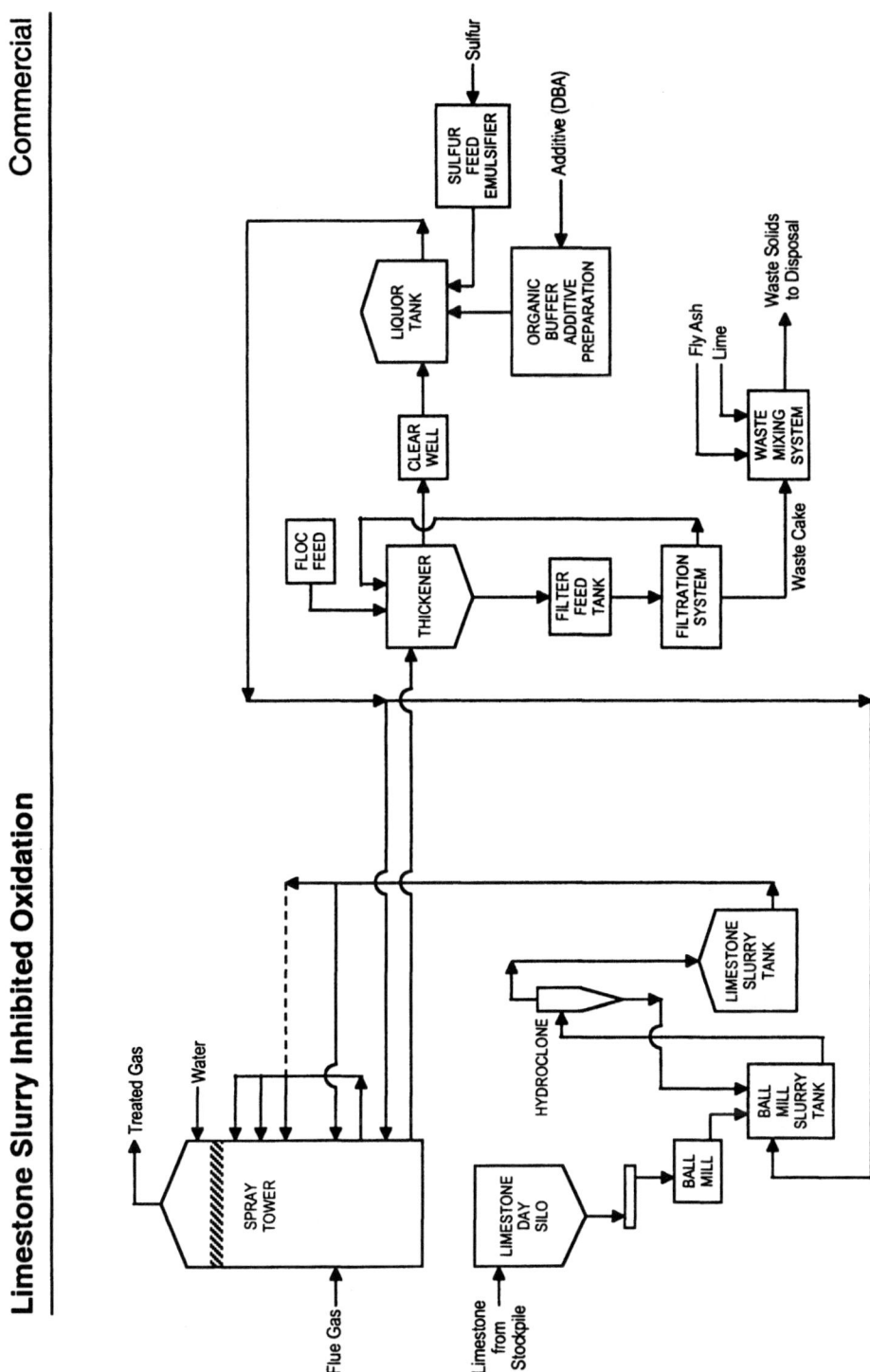

Limestone Slurry – Magnesium-Promoted

Commercial
Absorption of SO$_2$ in limestone slurry containing dissolved magnesium salts, followed by solids separation and usually admixture of the solids with fly ash and lime for disposal.

Process Characteristics
SO$_2$ Sorbent: A mixture of limestone and dissolved magnesium salts
Principal Raw Materials: Limestone; Dolomitic limestone
Potentially Saleable Byproducts: None
Solid Wastes: A mixed calcium sulfite and sulfate waste solids usually admixed with fly ash (and lime)
Liquid Effluents: None
Other Gaseous Emissions: None
Inlet SO$_2$ Levels: <1,000 to 2,500 ppm (Commercial Operations)
SO$_2$ Removal: Typical Design Range: 85-95%; Maximum at Highest Inlet SO$_2$: ~90%
NO$_X$ Removal: Nil - Although additives are under development that can be used for integrated control (e.g., soluble iron chelates)
Particulate Removal: Integrated control capability, but need for mixture of ash with FGD waste in some applications limits coupling of particulate control

Commercial Application
Number/Types: Several: Utility & industrial boilers
Locations: US
First Deployment: Late 1970s
Current Status: Inactive

Process Description
In a typical system, the gas contactor consists of a multi-level spray tower or deluge type device (e.g., venturi scrubber) or combination of both with an internal recirculation tank at the bottom. A slurry of limestone and magnesium hydroxide is fed to the recirculation tank which is recirculated through the contactor at a high L/G ratio. The presence of the magnesium increases the alkalinity of the solution enhancing SO$_2$ removal. Slurry is continuously withdrawn from the scrubber and dewatered usually by conventional thickening and filtration. In most systems on coal-fired boilers, the dewatered solids are usually admixed with fly ash and lime to improve handling and promote solidification for disposal because the solids typically cannot be dewatered as well as those produced in systems without magnesium addition. An option is to pump thickened solids directly to onsite ponds where the solids are settled out and the clarified liquor returned to the system, although this practice has been generally phased out. There are two approaches to providing the magnesium hydroxide used in the absorption. First, is use of dolomitic lime slaked onsite; and the second, is addition of Mg(OH)$_2$ slurry that can be separately added as a purchased slurry or from MgO slaked onsite.

Advantages/Disadvantages

Advantages
- Use of magnesium alkali addition increases the scrubbing alkalinity thereby enhancing SO$_2$ removal efficiency and reducing sulfate formation potential

Disadvantages/Limitations
- Produces a solid waste of little to no reuse potential
- Solid waste usually requires admixture with dry ash for disposal in landfills

Principal Suppliers
Bechtel   M. W. Kellogg/Weir

Waste Producing Processes 19

Limestone Slurry — Magnesium-Promoted Commercial

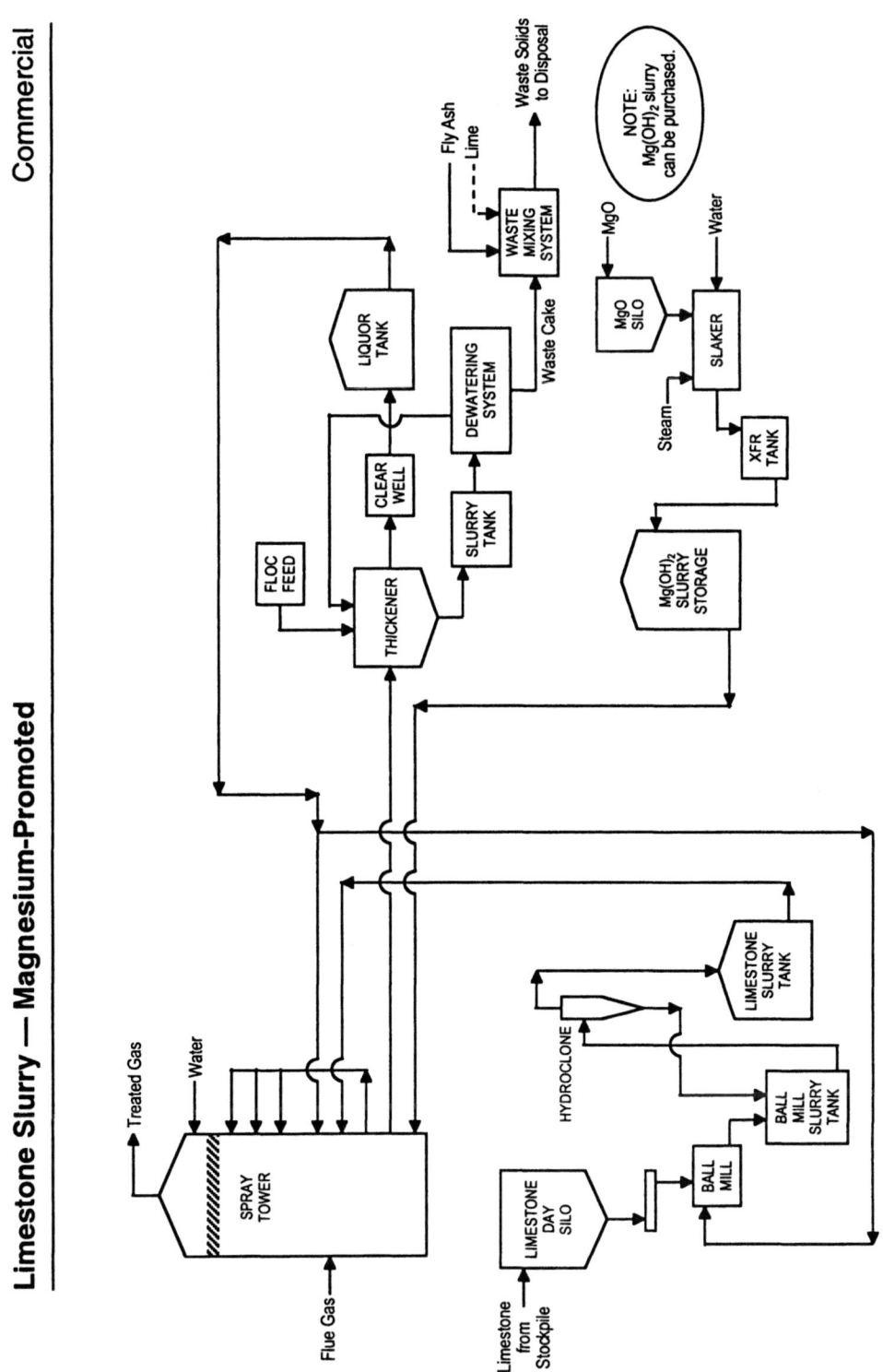

Alkaline Ash Scrubbing

Commercial

Absorption of SO_2 in an alkaline ash slurry sometimes supplemented with lime or limestone, followed by solids separation by thickening and filtration or ponding.

Process Characteristics
SO_2 Sorbent: Alkaline ash sometimes augmented with lime or limestone
Principal Raw Materials: Alkaline ash; (lime or limestone)
Potentially Saleable Byproducts: None
Solid Wastes: A mixed calcium sulfite and sulfate waste solids with fly ash
Liquid Effluents: None
Other Gaseous Emissions: None
Inlet SO_2 Levels: 600 to 2,000 ppm
SO_2 Removal: Typical Design Range: 60-95%
NO_x Removal: None
Particulate Removal: Integrated control capability

Commercial Application
Number/Types: Numerous: Utility boilers
Locations: US; Europe
First Deployment: Mid 1970s
Current Status: Active

Process Description
There are two major variations in alkaline ash scrubbing. The first involves the scrubbing approach – either combined particulate and SO_2 control or separate upstream, dry particulate control followed by SO_2 control. For combined control, a venturi scrubber is used followed by a "knockout" tower usually containing sprays and/or trays for additional SO_2 removal plus a demister. The trays/sprays may have common or separate recirculation loops with the venturi scrubber. Spent slurry withdrawn from the scrubber system is then dewatered using a ponding system or mechanical thickeners and filters, with recycle of the clarified water to the scrubbing system. While combined particulate and SO_2 control is lower capital cost than separate control, the biggest drawback is greater difficulty in controlling absorption chemistry due to the uncontrolled addition of alkalinity from the alkaline ash. In separate upstream, dry particulate control, the ash is collected and stored in a silo and then metered to the scrubber on demand, allowing much better chemistry control. Combined quench/spray towers are generally employed, but venturis are also used. The second major variation is use of supplemental alkali – calcitic lime (high CaO content); dolomitic lime (high magnesium content lime); or, in a few cases limestone. Almost every alkaline ash system is designed for the use of some supplemental alkali; however, only about half of the 16 units at seven plants in the US actually routinely use supplemental alkali. It is also noteworthy that almost all alkaline ash scrubbing systems in the US incorporate flue gas reheat, primarily because these are older, western units with plume visibility regulatory constraints. Flue gas reheat is rarely used in new (eastern) US plants.

Advantages/Disadvantages

Advantages
- Combined particulate and SO_2 removal capability
- Minimizes or eliminates sorbent raw materials

Disadvantages/Limitations
- Produces a solid waste of little to no reuse potential
- Many systems have experienced scaling and plugging problems due to the difficulty in controlling chemistry upsets
- Only applicable to highly alkaline ash coals

Principal Suppliers
Bechtel   M. W. Kellogg/Weir   Thyssen/CEA

Waste Producing Processes

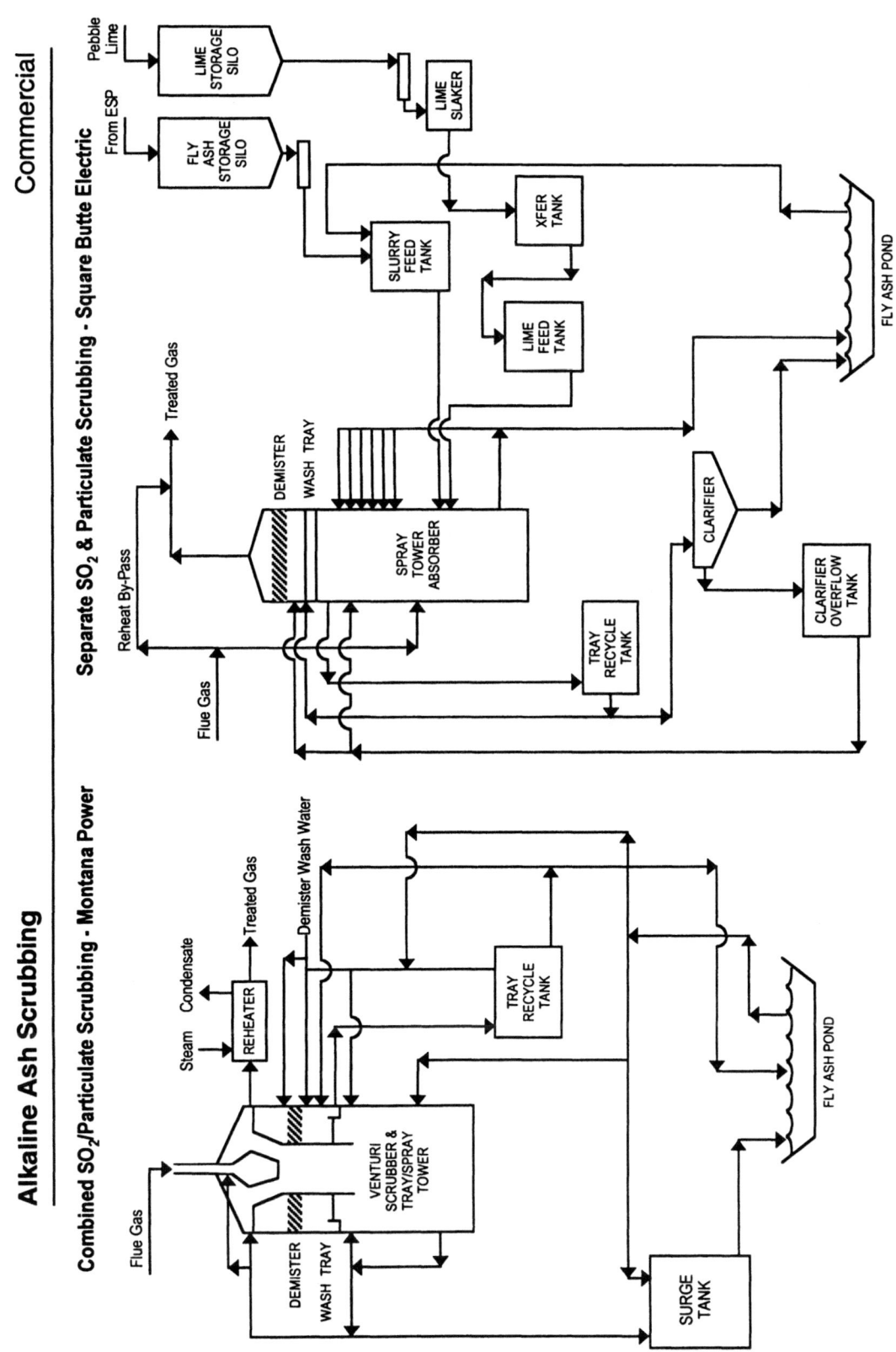

Sodium/Lime Dual Alkali (Concentrated Mode)

Commercial
Absorption of SO_2 in sodium sulfite solution, reaction of spent liquor with lime to regenerate absorbent producing waste solids for removal by thickening/filtration.

Process Characteristics
SO_2 Sorbent: Sodium sulfite/bisulfite or hydroxide/sulfite solution
Principal Raw Materials: Lime (hydrated or quick lime); Caustic or soda ash
Potentially Saleable Byproducts: None
Solid Wastes: Mixed calcium sulfite/sulfate wastes
Liquid Effluents: None
Added Gaseous Emissions: None
Inlet SO_2 Levels: 1,200 to 150,000 ppm (Commercial Operations)
SO_2 Removal: Typical Design Range: 90-99.5%; Maximum at Highest Inlet SO_2: 99.9+%
NO_x Removal: Nil
Particulate Removal: Integrated control capability

Commercial Application
Number/Types: Many: Utility & industrial boilers; coke furnaces; ore roasters & furnaces; chemical process units
Locations: North America; Europe; Far East; India
First Deployment: 1970s
Current Status: Active

Process Description
Flue gas is contacted (usually) in a tray tower, packed tower or disk/doughnut tower with a solution of sodium sulfite/bisulfite or sodium sulfite/hydroxide which absorbs the SO_2. Spent solution of sodium sulfite/bisulfite is continuously withdrawn to a reactor system where it is contacted with hydrated lime. The lime regenerates the absorbent solution and precipitates a mix of calcium sulfite solids containing a small amount of co-precipitated calcium sulfate. Solids are removed by thickening and filtration, and regenerated solution is returned to the scrubber. Filter cake is usually washed to recover sodium salts. Sodium makeup varies from 2% to 10% of the absorbed SO_2 depending upon the solution concentration used and the degree of cake wash. Oxidation of absorbent to sulfate is managed by the co-precipitation of calcium sulfate and sodium sulfate losses in the cake.

Advantages/Disadvantages

Advantages
- Clear solution scrubbing minimizes scaling and plugging problems
- High removal efficiencies and tolerant of wide fluctuations in inlet SO_2
- Wastes can be disposed without ash admixture
- Capable of integrated SO_2 and particulate control

Disadvantages/Limitations
- Produces a waste solid of little/no reuse potential
- Generally not applicable to very low sulfur applications due to relatively higher capital costs required to handle the increased oxidation of absorbent

Principal Suppliers
Advanced Air Technology	Anderson 2000	Marsulex
AirPol/FLS Miljo	Arthur D. Little	Ontario Hydro

Waste Producing Processes

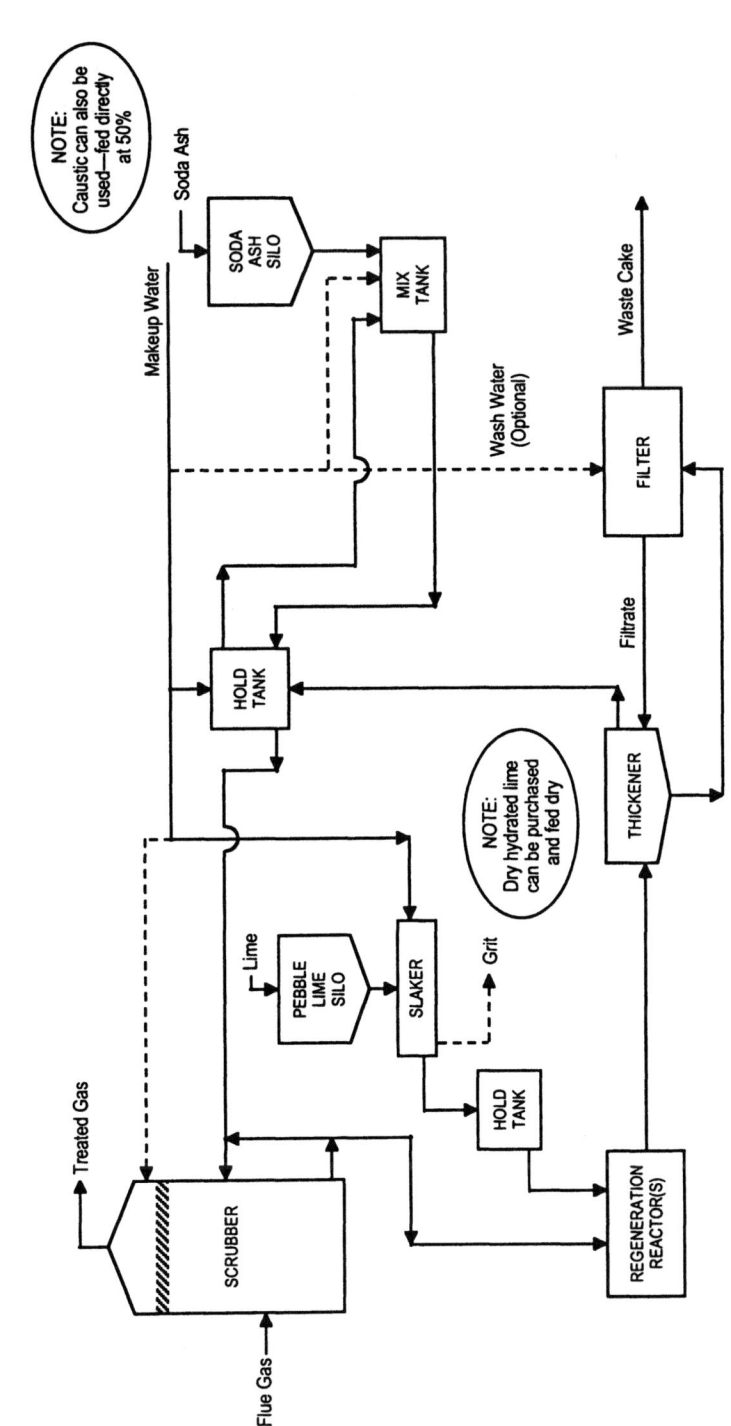

Sodium/Lime Dual Alkali (Concentrated Mode) — Commercial

Lime-Based Spray Dry Absorption (SDA)

Commercial
Flue gas is contacted in a spray dryer with slurry of lime and recycled solids. The gas is partly humidified and SO_2 is absorbed. Solids are collected in a fabric filter or ESP.

Process Characteristics
SO_2 Sorbent: Lime
Principal Raw Materials: Lime; Sodium chloride (sometimes required)
Potentially Saleable Byproducts: None
Solid Wastes: Mixed lime, calcium sulfite/sulfate solids and ash
Liquid Effluents: None
Added Gaseous Emissions: None
Inlet SO_2 Levels: <100 to ~3,000 ppm (Commercial Operations); ~3,500 ppm (Demonstration Testing)
SO_2 Removal: Typical Design Range: 80-93%; Maximum Achievable at Highest SO_2: ~95%
NO_x Removal: Nil
Particulate Removal: Integrated control capability

Commercial Application
Number/Types: Many: Utility and industrial boilers; waste incinerators; metallurgical plants; refineries; pulp and paper
Locations: Global
First Deployment: Early 1980s
Current Status: Active

Process Description
Flue gas is contacted and partially humidified in a spray dryer with a slurry of lime, recycled calcium sulfite/sulfate and ash. The spray dryer discharges to either a fabric filter or electrostatic precipitator where the solids are collected. A portion of the solids are recycled and slurried with fresh lime. The remaining waste solids are conveyed to a storage silo from which they are loaded directly into trucks or railcars for disposal. In general, lime-based SDA processes are constrained by limitations on water evaporation capacity. The amount of water that can be fed to the spray dryer (thus lime slurry feed) presents a practical limit for most spray dryer systems in most applications to about 3,000 ppm at 90-95% removal, although the theoretical limit is more nearly 4,000 ppm at 90% removal. Facilities with higher flue gas temperatures (more evaporative capacity) allow operation at higher inlet SO_2 levels. The presence of chloride increases SO_2 removal and reduces lime consumption especially at high SO_2 levels. Use of silica treated lime (the ADVACATE process) is also reported to increase removal and lime utilization. The principal differentiating factor in spray dryers is the manner of spray atomization - dual fluid nozzles versus rotary atomizers. Rotary atomizers are preferred for higher SO_2 levels.

Advantages/Disadvantages

Advantages
- Integrated SO_2 and particulate control
- Relatively low maintenance and operating labor
- Capable of high turndown by recycling clean gas
- Generally high SO_3 removal rates
- Partially saturated stack gas requires no reheat

Disadvantages/Limitations
- Produces a waste solid of little reuse potential
- Limitations on SO_2 removal at high inlet SO_2
- Generally high lime requirements
- Problems with production of "sticky" solids in installations ammonia injection used in SCR and SNCR for NO_x control

Principal Suppliers

ABB/EPA (ADVACATE)	B&W	Niro	Wheelabrator
AirPol/FLS Miljo (GSA)	Belco/Societe LAB	Procedair	
Amerex	Environmental Elements	Research Cottrell	

Waste Producing Processes

Lime-Based Spray Dry Absorption — Commercial

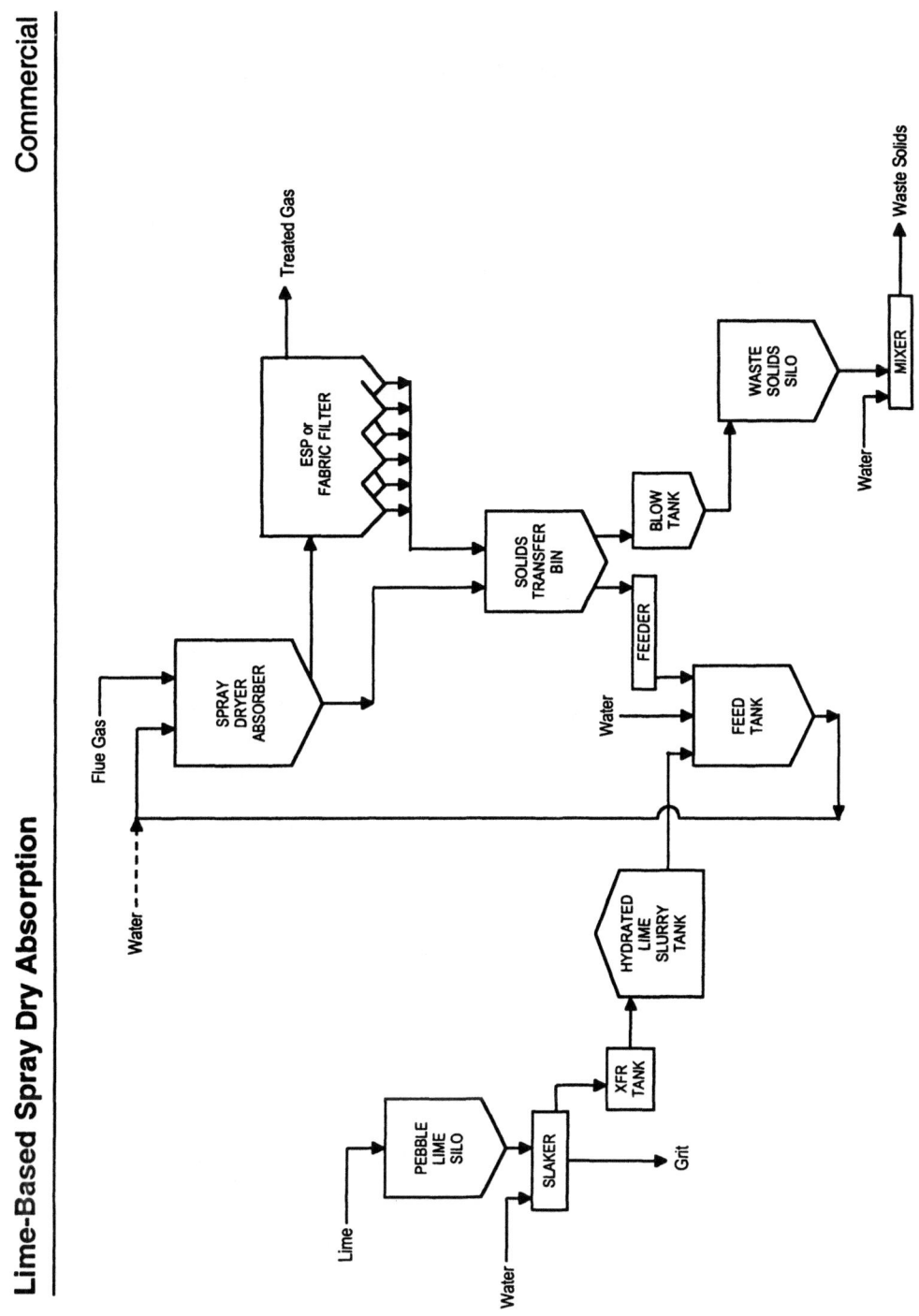

Sodium-Based Spray Dry Absorption (SDA)

Commercial
Flue gas is contacted in a spray dryer with a sodium salt slurry/solution, partially humidifying the gas and absorbing SO_2. Waste solids are collected downstream.

Process Characteristics
SO_2 Sorbent: Sodium alkali
Principal Raw Materials: Sodium alkali (Soda ash, nacholite, trona, waste brine)
Potentially Saleable Byproducts: None
Solid Wastes: Mixed sodium alkali, sodium sulfite/sulfate solids and ash
Liquid Effluents: None
Added Gaseous Emissions: None
Inlet SO_2 Levels: <100 to 1,800 ppm (Commercial Operations)
SO_2 Removal: Typical Design Range: 85-95%; Maximum Achievable at Highest SO_2: ~98%
NO_x Removal: Nil
Particulate Removal: Integrated control capability

Commercial Application
Number/Types: Several: Mostly utility and industrial boilers; general process operations
Locations: North America; Europe; Asia
First Deployment: Early 1980s
Current Status: Active

Process Description
Flue gas is contacted in a spray dryer with a slurry of sodium alkali salt, recycled sodium sulfite/sulfate solids and ash. The spray dryer discharges to either a fabric filter or electrostatic precipitator where the solids are collected. A portion of the solids may be recycled and slurried with fresh sodium alkali feed. The remaining waste solids are conveyed to a storage silo from which they are loaded directly into trucks or railcars for disposal. In general, SDA processes are constrained by limitations on water evaporation capacity of the gas which limits the amount of water that can be fed to the spray dryer and thereby the concentration of feed solids, but sodium-based SDAs are limited to a lesser extent than lime-based spray dryer systems. The practical limit for most spray dryer systems in fossil fuel combustion system applications is 3,000 ppm at 90-95% removal, although the theoretical limit is more nearly 4,000 ppm at 90% removal. Sodium based systems are probably capable of 20-25% higher limits than lime-based systems. Facilities with higher flue gas temperatures (more evaporation) may allow operation at even higher inlet SO_2 levels. The biggest drawbacks to sodium-based spray dryers are the high cost of the sodium makeup and disposal of a waste with a high solubles.

Advantages/Disadvantages

Advantages
- Integrated SO_2 and particulate control
- Relatively low maintenance and operating labor
- Little to no scaling or plugging potential
- Capable of high turndown by recycling clean gas
- Generally high SO_3 removal rates
- Partially saturated stack gas requires no reheat

Disadvantages/Limitations
- Produces a waste solid of little to no reuse potential and which can be troublesome to dispose because of the sodium salts solubility
- Limitations on SO_2 removal at high inlet SO_2
- Generally sodium cost if commercial alkali
- Problems with production of "sticky" solids in installations ammonia injection used in SCR and SNCR for NO_x control

Principal Suppliers
ABB	Belco/Societe LAB	UOP
B&W	Procedair	Wheelabrator

Waste Producing Processes

Sodium-Based Spray Dry Absorption — Commercial

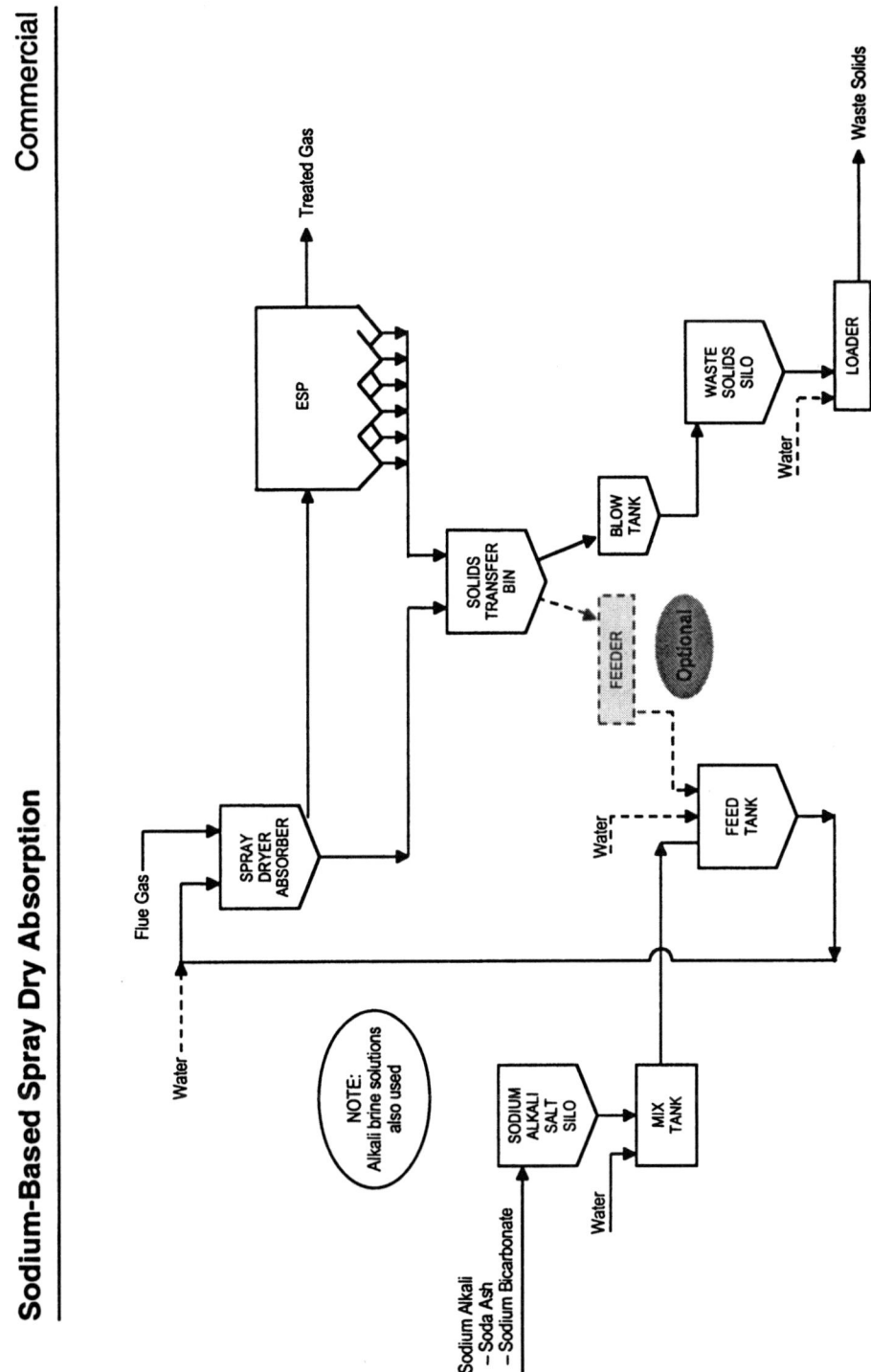

Circulating Fluid/Entrained Bed – Lime-Based

Commercial
Partial humidification of the gas, then sorption of SO_2 in an entrained bed of lime, recycled calcium-sulfur solids and ash, which are collected in a fabric filter or ESP.

Process Characteristics
SO_2 Sorbent: Lime
Principal Raw Materials: Lime; Sodium chloride (sometimes required)
Potentially Saleable Byproducts: None
Solid Wastes: Mixed lime, calcium sulfite/sulfate solids and ash
Liquid Effluents: None
Added Gaseous Emissions: None
Inlet SO_2 Levels: <100 to 3,800 ppm (Commercial Operations); up to 6,500 ppm (Demonstration Testing)
SO_2 Removal: Typical Design Range: 85-95%; Maximum Achievable at Highest SO_2: ~97%
NO_x Removal: Nil
Particulate Removal: Integrated control capability

Commercial Application
Number/Types: Many: Utility and industrial boilers; waste incinerators
Locations: US; Europe
First Deployment: Late 1980s
Current Status: Active

Process Description
Flue gas is partially humidified in a venturi/spray section and then contacted with a mix of hydrated lime, calcium sulfite/sulfate solids and ash in an up flow, cocurrent fluid bed-type reactor. Solids entrained in the gas flow out the top of the contactor to either a fabric filter or electrostatic precipitator where the solids are collected. A portion of the solids are recycled, mixed with lime and fed dry to the contactor. Waste solids are conveyed to a storage silo from which they are loaded directly into trucks or railcars for disposal. Since hydrated lime is fed dry, either hydrated lime can be purchased or pebble lime slaked onsite in a dry lime hydrator. Dry lime feed allows applications at much higher inlet SO_2 and removal efficiencies than Spray Dry Absorption systems. For high inlet SO_2 inlet levels and high removal efficiencies, the presence of chloride enhances SO_2 removal and reduces lime requirements.

Advantages/Disadvantages

Advantages
- Integrated SO_2 and particulate control
- Low maintenance and operating labor
- Little to no scaling or plugging potential - no handling of slurries or dewatering systems
- High removal efficiencies for other acid gases, especially SO_3, HCl, and HF
- Can operate on higher inlet SO_2 levels and removal efficiencies than Spray Dry Absorption
- Capable of high turndown by recycling clean gas
- Partially saturated stack gas requires no reheat

Disadvantages/Limitations
- Produces a waste solid of little/no reuse potential
- Lime consumption is relatively high, comparable to Spray Dry Absorption systems
- Problems with production of "sticky" solids in installations ammonia injection used in SCR and SNCR for control

Principal Suppliers

ABB	Amerex	Lentjes-Bischoff (Lurgi)
AirPol/FLS Miljo (GSA operated on dry lime feed)	Environmental Elements (Lentjes-Bischoff/Lurgi)	Procedair
		Wulff Gmbh

Waste Producing Processes

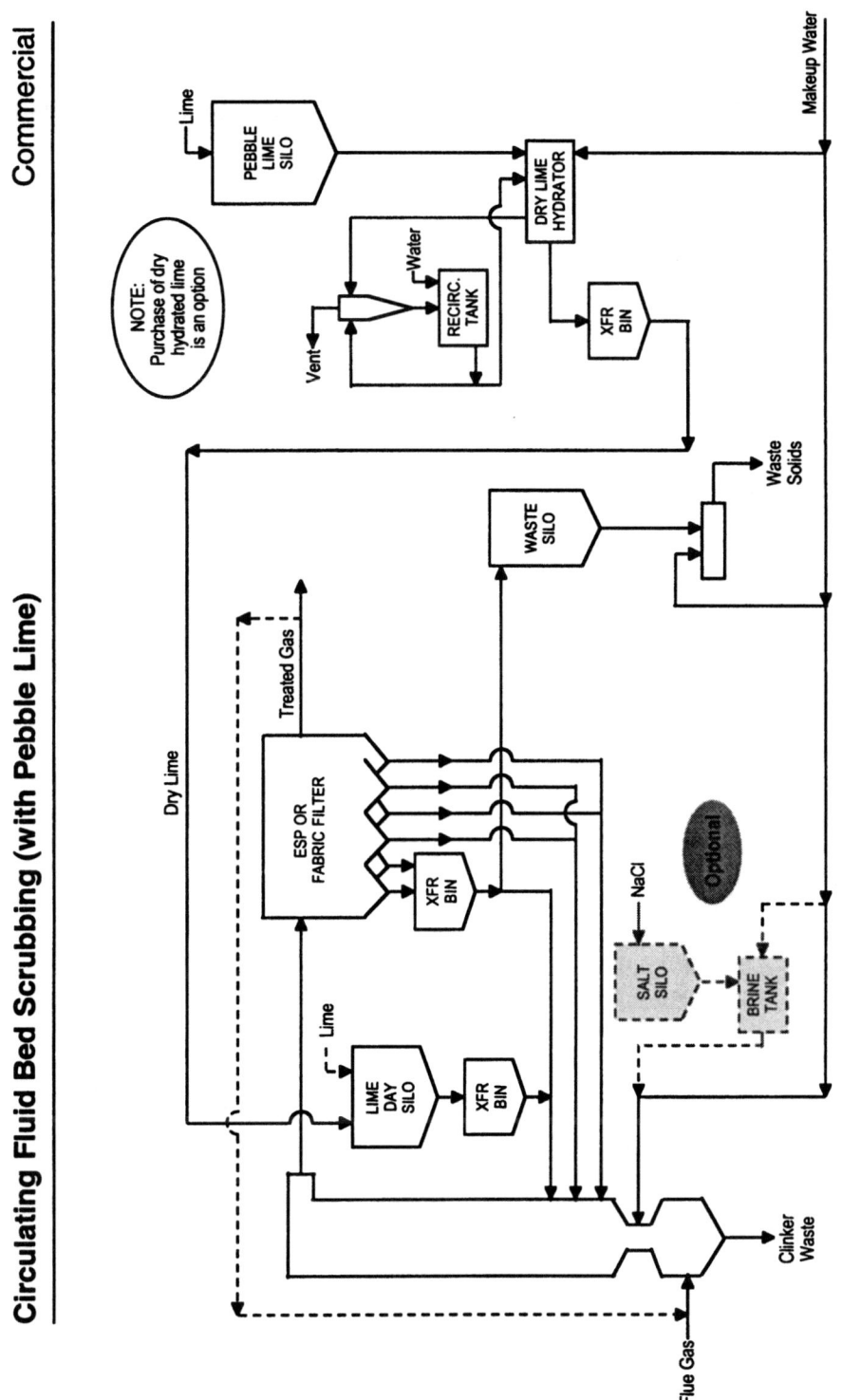

Sodium-Based Duct Injection

Commercial
A sodium alkali sorbent is injected dry into the flue gas duct upstream of a particulate collector. Solids are collected downstream in a fabric filter or ESP. SO$_2$ removal is effected in the duct and particulate collector.

Process Characteristics
SO$_2$ Sorbent: Sodium alkali
Principal Raw Materials: Sodium alkali salt (e.g., soda ash, sodium bicarbonate, nacholite)
Potentially Saleable Byproducts: None
Solid Wastes: Mixed sodium sulfite/sulfate solids and ash
Liquid Effluents: None
Added Gaseous Emissions: None
Inlet SO$_2$ Levels: 300 to 1,200 ppm (Commercial Operations)
SO$_2$ Removal: 25-80% (Commercial Operations) – Highly dependent upon conditions
NO$_x$ Removal: Nil
Particulate Removal: Integrated control capability

Commercial Application
Number/Types: Numerous: combustion boilers; MSW incinerators; refineries; process units
Locations: US; Europe
First Deployment: Late 1980s
Current Status: Active

Process Description
Sodium alkali is injected into the flue gas ductwork upstream of a particulate collector (fabric filter or electrostatic precipitator). SO$_2$ removal is effected during contact in both the ducting and the particulate collector. There are a number of processes that vary primarily in terms of the type of sodium alkali used; sorbent preparation; the manner of injection (wet versus dry); duct residence time (longer helps); and type of particulate collector. The degree of humidification of the gas is one of the most important factors. Totally dry systems do not achieve the levels of SO$_2$ removal or alkali utilization as those using slurry injection or water sprays to partially humidify the gas. A fabric filter is also generally better than an ESP because of the additional gas/sorbent contact time provided on the bag surface. In this regard, fabric filter cleaning cycles have been shown to have an affect on SO$_2$ removal and alkali utilization. Sodium salts used include sodium carbonate, sodium bicarbonate, sodium sesquicarbonate, refined trona (naturally occurring sodium carbonate and bicarbonate) and nacholite (naturally occurring sodium bicarbonate). Most successful have been sodium bicarbonate and nacholite in terms of both SO$_2$ removal and salt utilization. This prompted establishment of NaTec to garner nacholite sources and market nacholite-based technology. Urea is also injected in some systems to suppress the formation of NO$_2$ during absorption that can result in an attached plume.

Advantages/Disadvantages

Advantages
- Integrated SO$_2$ and particulate control
- Relatively low maintenance and operating labor
- Capability for high removal of other acidic species including SO$_3$ and HCl
- Partially saturated stack gas requires no reheat

Disadvantages/Limitations
- Produces a waste solid of little reuse potential
- Limitations on SO$_2$ removal at high inlet SO$_2$
- Potential for high solubles in the waste may constrain disposal alternatives

Principal Suppliers
ABB	EPRI	Many Proprietary	Wheelabrator
Airborne (formerly NaTec)	Solvay	("Home Grown")	

Waste Producing Processes

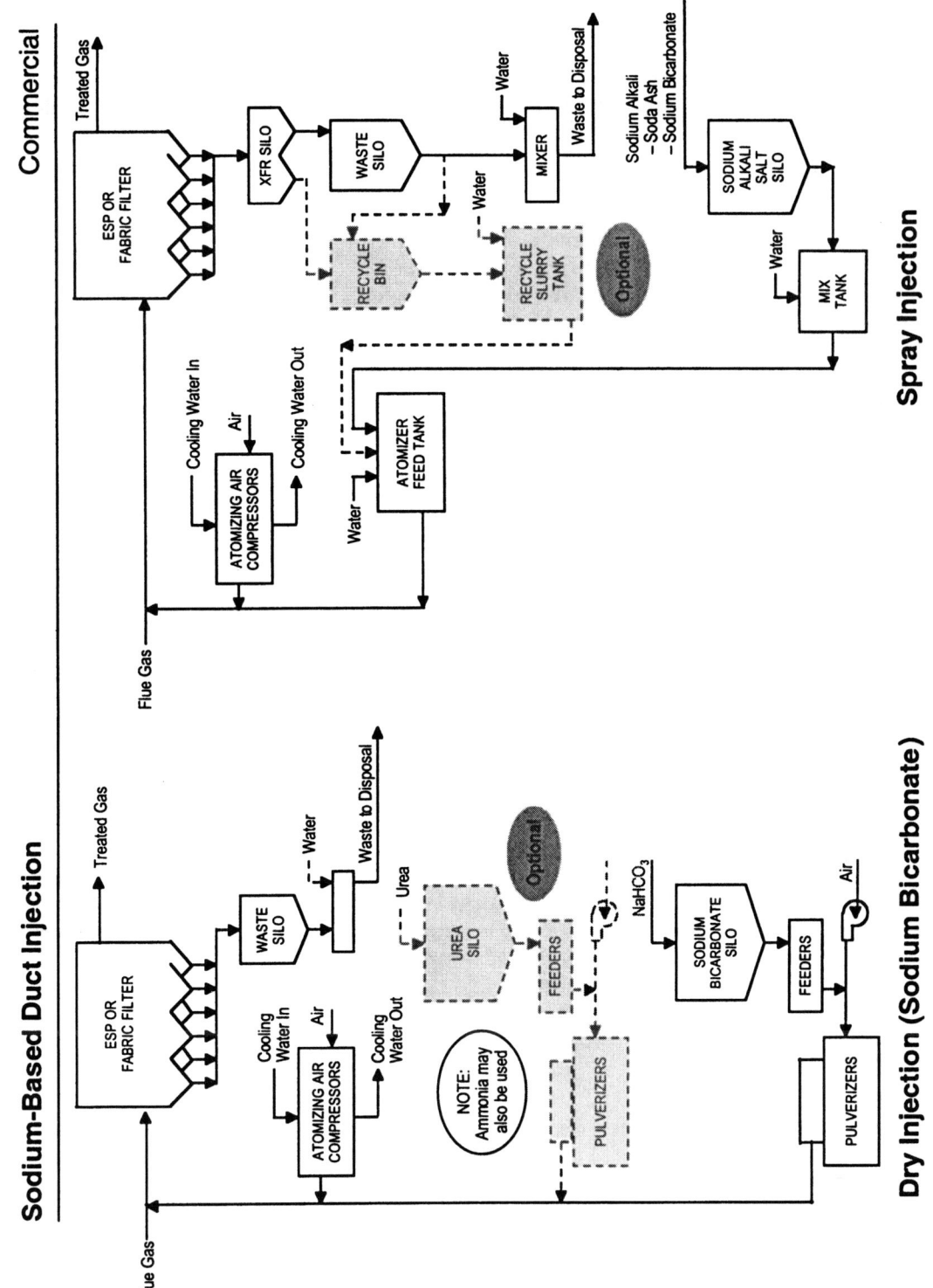

Furnace Sorbent Injection

Commercial
Limestone (or lime) is injected into the boiler sometimes coupled with downstream water spray humidifiers. Solids are collected in a fabric filter or ESP. SO_2 is absorbed in the boiler, duct and collector.

Process Characteristics
SO_2 Sorbent: Lime (CaO)
Principal Raw Materials: Lime or limestone
Potentially Saleable Byproducts: None
Solid Wastes: Mixed lime, calcium sulfate solids and ash
Liquid Effluents: None
Added Gaseous Emissions: None
Inlet SO_2 Levels: 700 to ~2,500 ppm (Tested)
SO_2 Removal: 45-90+% (Tested) – Highly dependent upon system and operating variables
NO_x Removal: Several designs include the capability for NO_x control (50-80+% possible)
Particulate Removal: Integrated control capability

Commercial Application
Number/Types: Numerous: Mostly utility and industrial boilers
Locations: Europe; US; Asia
First Deployment: Late 1980s (Current technology)
Current Status: Active

Process Description
Limestone or lime is injected into the combustion furnace (usually a wall-fired or tangentially-fired boiler) in an optimal temperature region for SO_2 absorption, usually in the range of 900-1200 C. Lime (CaO) or limestone calcined by the high combustion temperatures reacts with the SO_2 forming anhydrous $CaSO_4$. Solids are then collected in a downstream particulate collector (fabric filter or ESP). There are numerous variations of the technology. These include: use of limestone or lime as noted; use of water spray humidifcation upstream of the particulate collector; recycle of partially absorbents; use of additives (principally sodium alkali); and the inclusion of combined NO_x control, usually through the injection of urea. SO_2 removal efficiency is highly mass transfer limited. Therefore, many variables impact performance: SO_2 concentration (lower is easier); type of lime/limestone (magnesium can help); degree of gas humidification (higher humidification helps); sorbent particle size (grinding helps); additives; duct contact time (longer helps); alkali feed stoichiometry (higher is better); and type of particulate collector (a fabric filter is usually better).

Advantages/Disadvantages

Advantages
- Integrated SO_2 and particulate control (and some with NO_x)
- Relatively low maintenance and operating labor
- Little to no scaling or plugging potential
- Stack gas requires no reheat

Disadvantages/Limitations
- Produces a waste solid of little reuse potential
- Generally limited to <90% SO_2 removal at high inlet SO_2 unless configured with downstream humidification that mimics spray dryer systems
- Generally high lime/limestone requirements

Principal Suppliers

Suppliers of Commercial Technology

ABB	Osterreichische
B&W (LIDS)	Draukraftwerke
Inland Steel/Research	Riley
Cottrell	Tampella (LIFAC)

Technology Developers

ABB/EPA (LIMB-ADVACATE)	Fossil Energy Research
B&W (ALIDS)	Steinmuller
Babcock Hitachi	TAV Trocken

Waste Producing Processes

Furnace Sorbent Injection — Commercial

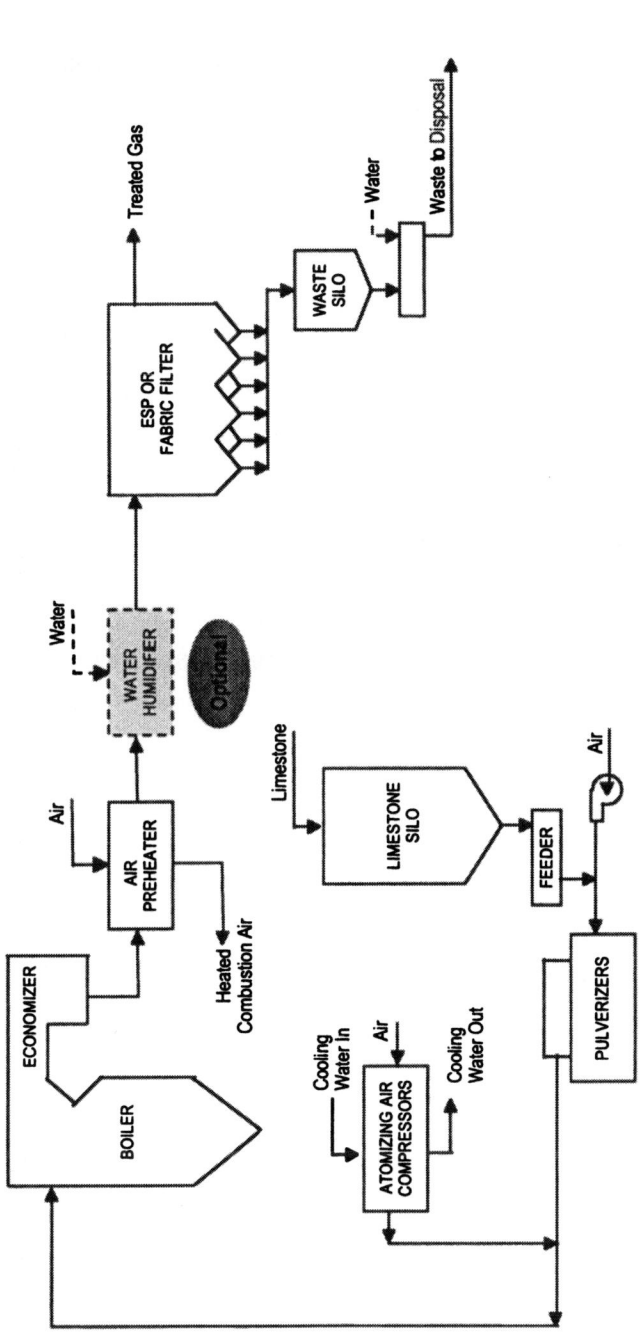

Sodium Solution – Once-Through

Commercial
Absorption of SO$_2$ in an alkaline sodium solution that is usually neutralized and oxidized to sodium sulfate prior to discharge.

Process Characteristics
SO$_2$ Sorbent: Sodium alkali solution (caustic, soda ash, sodium bicarbonate)
Principal Raw Materials: Sodium alkali
Potentially Saleable Byproducts: None (In Japan, Sodium sulfite or sulfate was occasionally marketed)
Solid Wastes: None
Liquid Effluents: Sodium sulfite or sulfate solution
Added Gaseous Emissions: None
Inlet SO$_2$ Levels: <100 to 100,000+ ppm
SO$_2$ Removal: Typical Design Range: 90-98%; Maximum Achievable at Highest SO$_2$: 99.9+%
NO$_x$ Removal: Limited NO$_x$ control in low sulfur applications; additives also used (e.g., ClO$_2$ in Sumitomo-Fujikasui)
Particulate Removal: Integrated control capability

Commercial Application
Number/Types: Many: Utility and industrial boilers; refinery boilers and process units; metallurgical plants; pulp and paper plants; chemical process units; general manufacturing units
Locations: Global
First Deployment: 1930s
Current Status: Active

Process Description
Once-through sodium solution scrubbing is one of the simplest forms of FGD technology involving scrubbing with a sodium alkali salt solution followed by minimal post-treatment prior to discharge (or occasionally reuse). Caustic, soda ash, sodium bicarbonate, trona (impure soda ash) and waste brines are all used. Virtually every type of scrubber vessel has also been employed. For high efficiency removal at high SO$_2$ levels, tray towers are favored. Venturis are common for combined particulate and SO$_2$ control. In large scale systems, the liquor is often sent directly to evaporation ponds. In smaller industrial plants, the liquor is frequently sent to a wastewater treatment plant or discharged after neutralization and oxidation.

Advantages/Disadvantages

Advantages
- Clear solution scrubbing minimizes scaling and plugging problems
- High removal efficiencies for a wide range of inlet SO$_2$ levels
- Tolerant of wide fluctuations in inlet SO$_2$
- Capable of integrated SO$_2$ and particulate control
Very uncomplicated process
- Relatively benign liquid waste

Disadvantages/Limitations
- Produces a liquid waste that may present environmental constraints in many applications
- High cost of sodium alkali unless very low SO$_2$ levels or waste alkali sources available
- Waste liquor has little value unless coupled with a captive consumer such as a pulp and •
paper plant

Principal Suppliers

Advanced Air Technology	Anderson 2000	Exxon (ERE)	Sumitomo--Fujikasuito
Airborne	Arthur D. Little	Kureha	Tsukishima-Bahco
Airpol/FLS Miljo	Belco	Marsulex	UOP
American Air Filter	Clean Gas Systems	Procedair	Wheelabrator
Amerex	EEC	Showa-Denko	Zurn

Waste Producing Processes

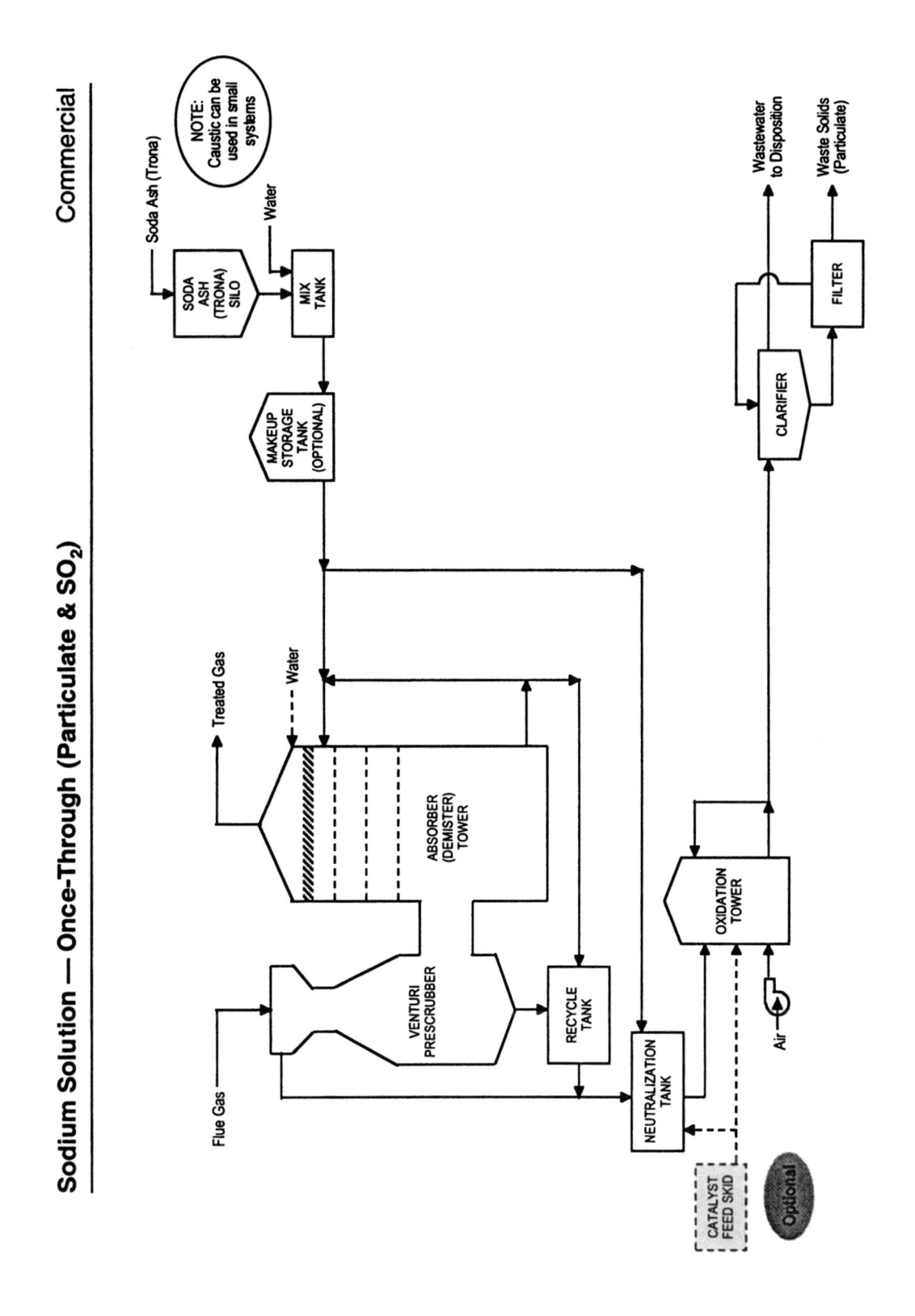

Sodium Solution — Once-Through (Particulate & SO$_2$)   Commercial

NOTE: Caustic can be used in small systems

Magnesium Hydroxide Solution – Once-Through

Commercial
Absorption of SO$_2$ in a slurry of magnesium hydroxide that is usually neutralized and oxidized to magnesium sulfate prior and discharged to a wastewater treatment plant. (In pulp plants magnesium salts can be recycled.)

Process Characteristics
SO$_2$ Sorbent: Magnesium sulfite/bisulfite – Solution and slurry
Principal Raw Materials: Magnesium oxide (slaked onsite) or magnesium hydroxide
Potentially Saleable Byproducts: Magnesium sulfite/bisulfite for captive use in magnesium-based pulp plants
Solid Wastes: None (unless combined with particulate control)
Liquid Effluents: Magnesium sulfate solution unless recycled as magnesium sulfite liquor
Added Gaseous Emissions: None
Inlet SO$_2$ Levels: up to 2,500 ppm
SO$_2$ Removal: Typical Design Range: 90%; Maximum Achievable at Highest SO$_2$: 98+%
NO$_x$ Removal: Nil
Particulate Removal: Integrated control capability

Commercial Application
Number/Types: Several: Utility and industrial oil-fired boilers; pulp and paper plant black liquor boilers
Locations: US; Far East
First Deployment: 1970s
Current Status: Active

Process Description
Once-through magnesium solution scrubbing is akin to once-through sodium scrubbing. It involves scrubbing with a recirculated solution of magnesium sulfite/bisulfite refreshed by addition of magnesium hydroxide to maintain operating pH. Open spray towers or towers with both sprays and open-type, self-draining type sieve trays are used because of the frequent presence of slurries either from Mg(OH)$_2$ or MgSO$_3$. There are two options for handling spent solution. One is reuse of the solution, which most commonly occurs in a captive pulp plant. The other is discharge through a post-treatment system involving neutralization with scrubber-bypassed magnesium hydroxide slurry followed by oxidation to more soluble magnesium sulfate. The solution is then sent to a wastewater treatment plant or discharged directly.

Advantages/Disadvantages

Advantages
- High removal efficiencies for a relatively wide range of inlet SO$_2$ levels
- Tolerant of wide fluctuations in inlet SO$_2$
- Capable of integrated SO$_2$ and particulate control
- Very uncomplicated process
- Relatively benign liquid waste

Disadvantages/Limitations
- Produces a liquid waste that may present environmental constraints in many applications if not reused (e.g., pulp plant)
- Waste liquor has little value unless coupled with a captive consumer such as a pulp and paper plant

Principal Suppliers

IHI	Marsulex	Taiwan Energy & Resources Laboratories
Kawasaki	Mitsui	Ube

Waste Producing Processes

Magnesium Hydroxide — Once-Through

Commercial

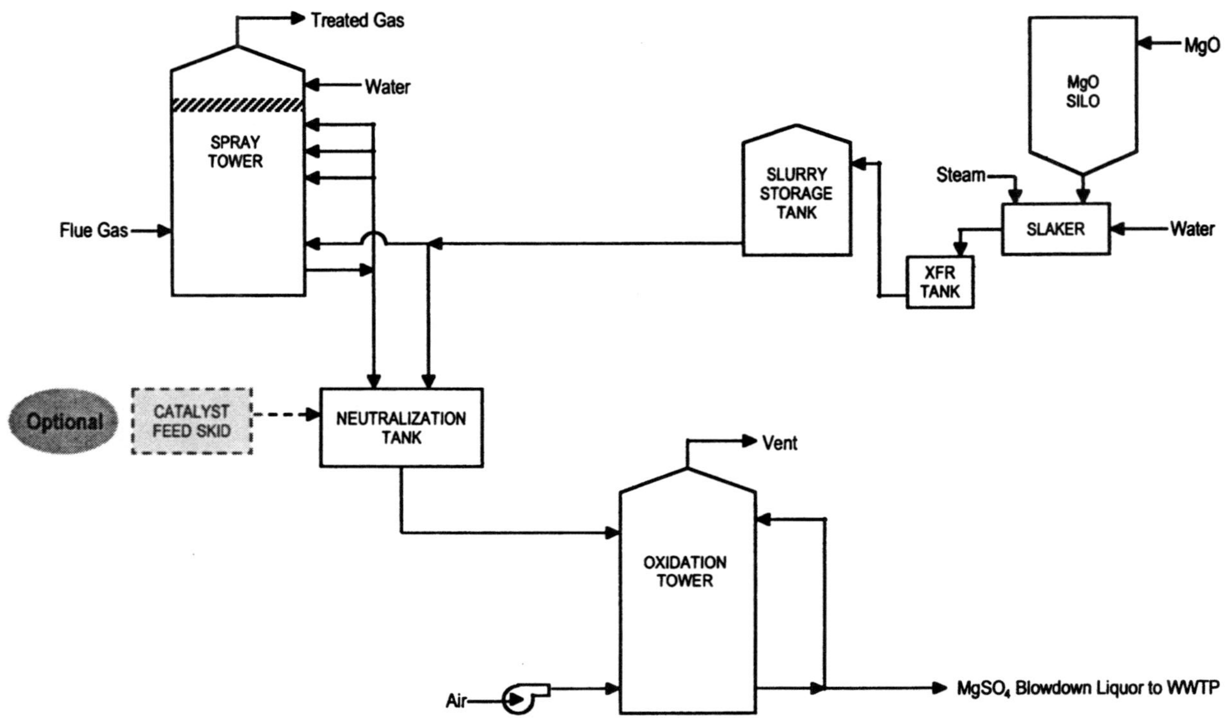

Seawater Scrubbing

Commercial

Absorption of SO$_2$ in an seawater, sometimes augmented with lime, followed by neutralization through mixture with additional seawater, addition of lime, or both and aeration prior to discharge.

Process Characteristics
SO$_2$ Sorbent: Natural seawater alkalinity (bicarbonate) sometimes augmented with lime
Principal Raw Materials: Seawater; (Lime)
Potentially Saleable Byproducts: None
Solid Wastes: None
Liquid Effluents: Spent neutralized seawater
Added Gaseous Emissions: None
Inlet SO$_2$ Levels: up to ~2,000 ppm (Commercial Operations)
SO$_2$ Removal: Typical Design Range: 80-95%; Maximum Achievable at Highest SO$_2$: 98+%
NO$_x$ Removal: Nil
Particulate Removal: Upstream decoupled particulate control needed for direct seawater discharge

Commercial Application
Number/Types: Many – Utility and industrial boilers; refineries
Locations: Scandinavia; India; Far East; South America
First Deployment: 1930s
Current Status: Active

Process Description
There are two basic approaches to seawater scrubbing and several variations on each. One approach uses only the natural alkalinity of the seawater; the other augments the natural alkalinity with addition of lime. Factors that play importantly in the design configuration are availability of seawater; inlet SO$_2$ level and removal requirement; and environmental constraints in discharging seawater. The simplest approach is once-through seawater scrubbing using a multiple contact tower (spray, packed or combination) with direct (return) discharge of the seawater possibly mixed with other seawater cooling water discharges. Variations in discharge treatment include forced oxidation of the scrubber discharge to convert essentially all sulfite/bisulfite to sulfate, pH adjustment usually with lime, and aeration to replenish a portion of the dissolved oxygen. Lime can also be added to the absorber to increase removal efficiency and decrease seawater requirement and L/G. Bechtel's version of seawater scrubbing involves use of a reaction tank external to the scrubber to react magnesium sulfate with lime producing magnesium hydroxide and precipitating gypsum. Mg(OH)2-rich solution is recirculated to the scrubber to enhance removal efficiency. The gypsum is redissolved in seawater prior to discharge.

Advantages/Disadvantages

Advantages
- Generally very low scale potentia
- Avoids solid waste production
- Highly economical if seawater is readily available, especially if it is used for plant CW

Disadvantages/Limitations
- Low SO$_2$ applications may be uneconomic due to low tolerance to oxidation
- Relatively complex and energy intensive
- High capital costs

Principal Suppliers

Flakt-Hydro/ABB
Bechtel

B&W
Lentjes-Bischoff (Bischoff)

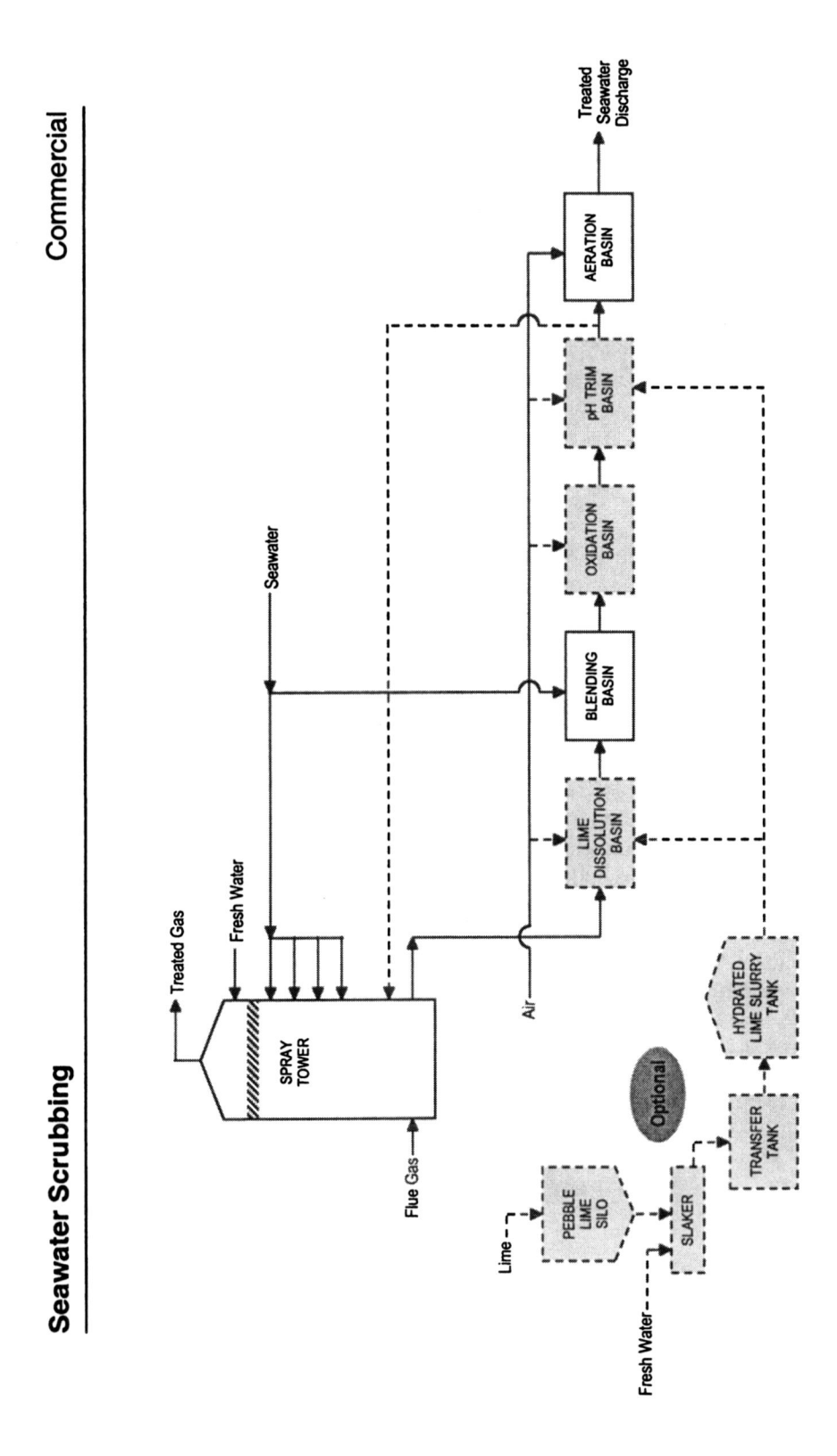

Waste Producing Processes

Seawater Scrubbing — Commercial

Sodium/Limestone Dual Alkali (Concentrated Mode)

Developmental
Absorption of SO_2 in a sodium sulfite/bisulfite solution, then reaction of spent liquor with limestone to regenerate the absorbent and produce a waste solid which is removed by thickening and filtration.

Process Characteristics
SO_2 Sorbent: Sodium sulfite/bisulfite solution
Principal Raw Materials: Limestone (constraints on type of limestone); Caustic or soda ash
Potentially Saleable Byproducts: None
Solid Wastes: Mixed calcium sulfite/sulfate wastes
Liquid Effluents: None
Added Gaseous Emissions: None
Inlet SO_2 Levels: 2,000 to 3,000 ppm (Tested)
SO_2 Removal: 90+% (Tested)
NO_x Removal: Nil
Particulate Removal: Integrated control capability

Commercial Application
Number/Types: None
Locations: NA
First Deployment: NA
Current Status: Inactive (Demonstration testing completed in 1980s)

Process Description
Flue gas is contacted in a tray tower or disk/doughnut tower with a solution of sodium sulfite/bisulfite which absorbs the SO_2. Spent solution of nearly pure bisulfite is continuously withdrawn to a multi-stage reactor system where it is contacted with limestone. The limestone regenerates the absorbent solution and precipitates a mix of calcium sulfite solids containing a small amount of co-precipitated calcium sulfate. Solids are removed by thickening and filtration, and regenerated solution is returned to the scrubber. Filter cake is usually washed to recover sodium salts. Sodium makeup rates are projected to be about 5% of the absorbed SO_2 depending upon the solution concentration used and the degree of cake wash. Oxidation of absorbent to sulfate is managed by the co-precipitation of calcium sulfate and sodium sulfate losses in the cake. Testing demonstrated the overall feasibility of the process, but two shortcomings have limited full commercialization. One is constraint on the type of limestone required (CaO content, crystal type and MgO content) and the other is the relatively high sensitivity of the process chemistry to upsets.

Advantages/Disadvantages

Advantages
- Clear solution scrubbing minimizes scaling and plugging problems
- Controlled precipitation reactions produce waste solids that can be disposed without admixture with ash
- Capable of integrated SO_2 and particulate control

Disadvantages/Limitations
- Produces a waste solid of little/no reuse potential
- Generally not applicable to very low sulfur applications due to relatively higher capital costs required to handle the increased oxidation of absorbent
- Requires a high quality limestone with constraints on crystal type and Mg content

Principal Developers
Arthur D. Little   Ontario Hydro (FMC)   US EPA

Waste Producing Processes

Sodium/Limestone Dual Alkali (Concentrated Mode)   Developmental

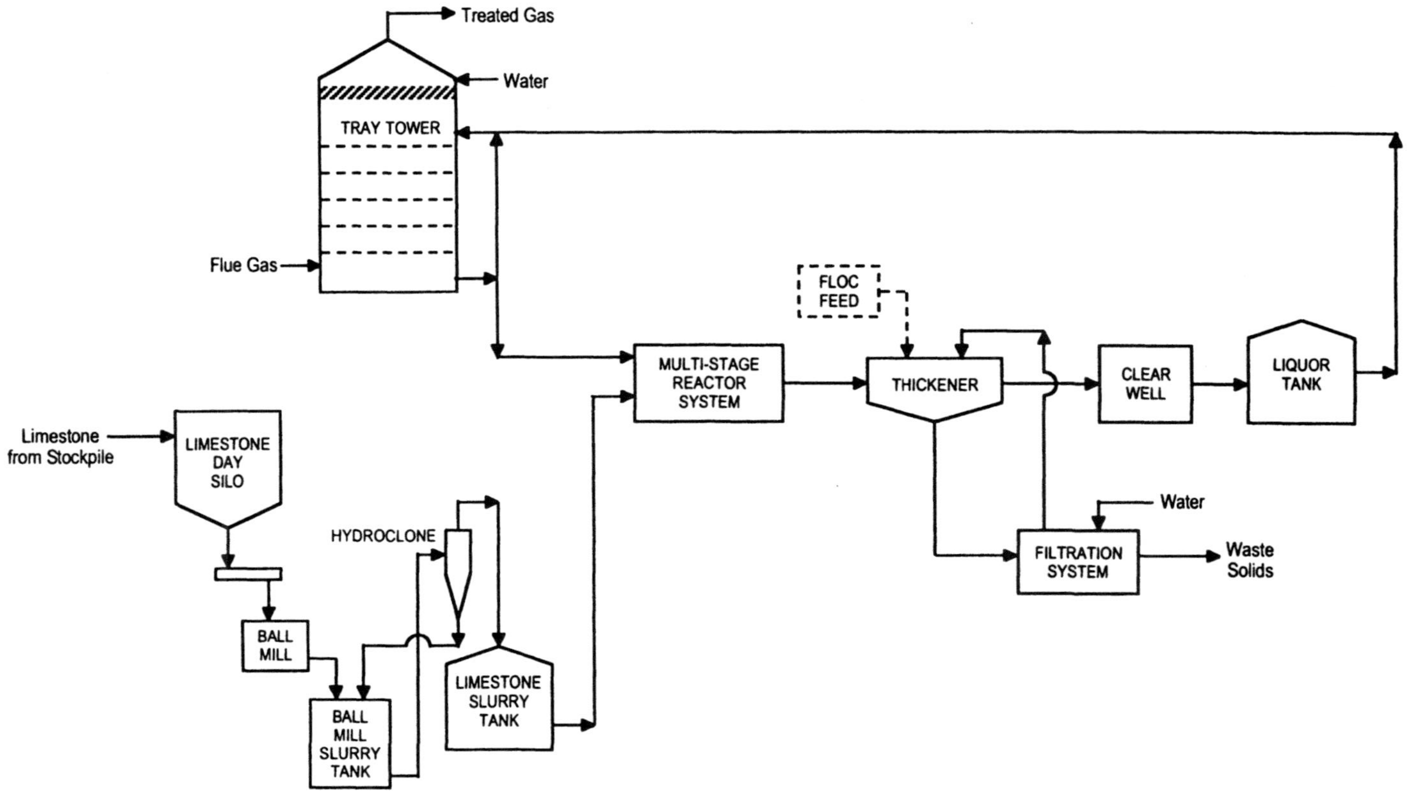

Circulating Fluid/Entrained Bed with NO$_x$ Control

Developmental
Injection of ammonia into hot flue gas, then contacting with an entrained bed of lime and catalyst (mixed with recycled calcium sulfite/sulfate solids and ash), which are collected in a fabric filter or electrostatic precipitator.

Process Characteristics
SO$_2$ Sorbent: Lime
Principal Raw Materials: Lime, Ammonia
Potentially Saleable Byproducts: None
Solid Wastes: Mixed lime, calcium sulfite/sulfate solids, iron sulfate and ash
Liquid Effluents: None
Added Gaseous Emissions: None
Inlet SO$_2$ Levels: up to 1,000 ppm (Tested)
SO$_2$ Removal: up to 97% (Tested)
NO$_x$ Removal: 80-95%
Particulate Removal: Integrated control capability

Commercial Application
Number/Types: None
Locations: NA
First Deployment: NA
Current Status: Active (In pilot testing)

Process Description
Hot flue gas, at a temperature of 600-850°F is injected with ammonia and then contacted with a mix of solids containing iron sulfate, hydrated lime, calcium sulfite/sulfate and ash in an up flow, cocurrent fluid bed-type reactor. The iron sulfate is an inexpensive NO$_x$ reduction catalyst and lime absorbs SO$_2$. The solids are entrained in the gas which flows out the top of the contactor to either a fabric filter or electrostatic precipitator where the solids are collected. A portion of the solids are recycled, mixed with lime and fed dry to the bottom of the contactor. Waste solids are conveyed to a storage silo from which they are loaded directly into trucks or railcars for disposal. Since hydrated lime is fed dry, either hydrated lime must be purchased or pebble lime slaked onsite in a dry lime hydrator. No water is used in the gas contacting system. For high inlet SO$_2$ inlet levels and high removal efficiencies, the presence of chloride enhances SO$_2$ removal and reduces lime requirements.

Advantages/Disadvantages

Advantages
- Integrated SO$_2$, particulate and NO$_x$ control
- Low maintenance and operating labor - little to no scaling or plugging potential
- No handling of slurries or dewatering systems
- High removal efficiencies for other acid gases, especially SO$_3$, HCl and HF
- Capable of high turndown by recycling clean gas
- Stack gas requires no reheat

Disadvantages/Limitations
- Produces a waste solid of little/no reuse potential
- Limitations on SO$_2$ removal at high inlet SO$_2$ but less so than Spray Dry Absorbers
- Generally high lime requirements as with Spray Dry Absorption systems

Principal Developers
Lentjes-Bischoff (Lurgi)

Waste Producing Processes

Circulating Fluid Bed with NO$_x$ Control

Developmental

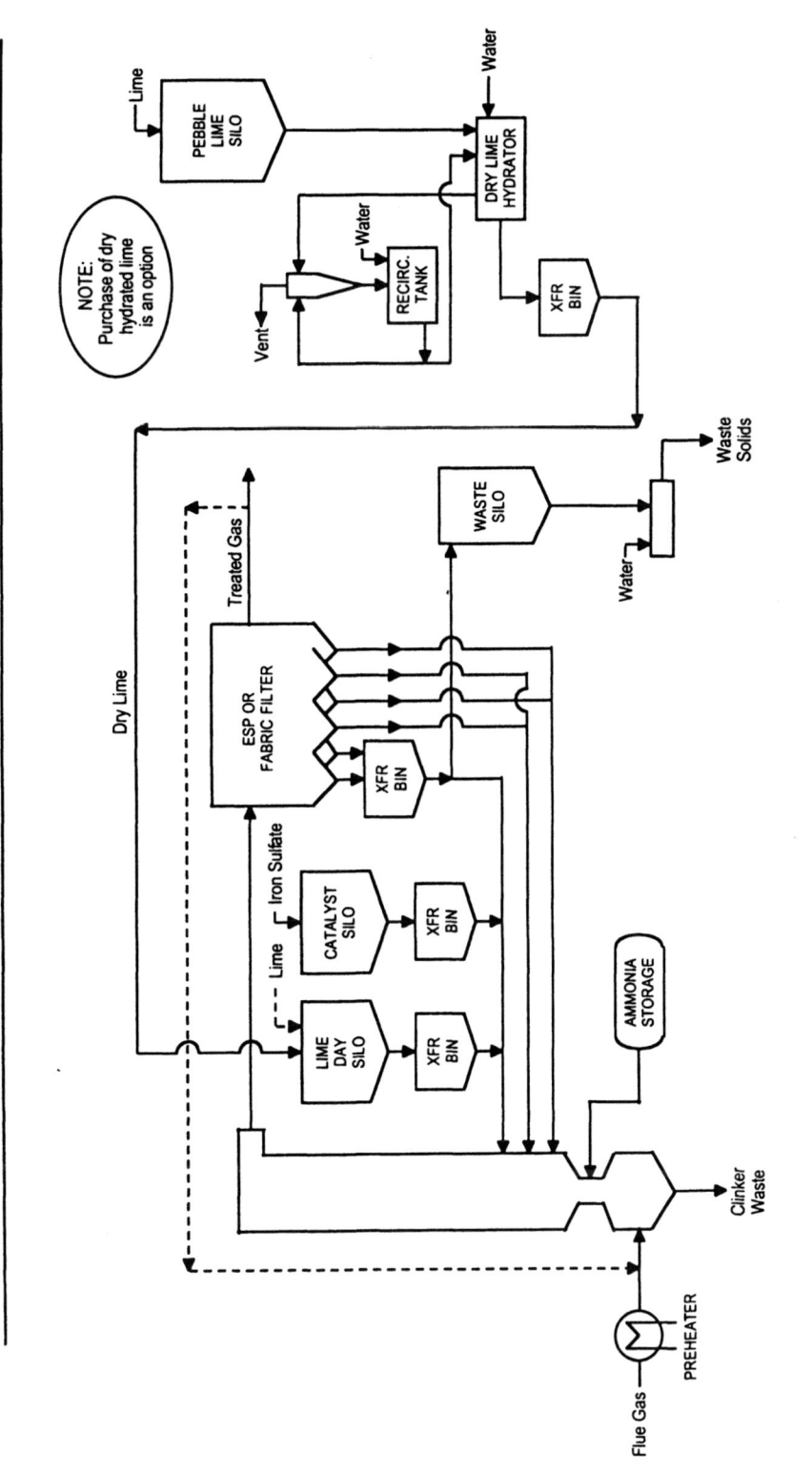

NOTE: Purchase of dry hydrated lime is an option

Lime-Based Duct Injection

Developmental
Lime is injected into the duct as slurry or dry with water spray humidification. Solids are collected in a fabric filter or ESP. SO_2 is absorbed in the duct and collector.

Process Characteristics
SO_2 Sorbent: Lime (typically with magnesium or sodium addition)
Principal Raw Materials: Lime (usually high magnesium content); Sodium (bi)carbonate (sometimes required)
Potentially Saleable Byproducts: None
Solid Wastes: Mixed lime, calcium sulfite/sulfate solids and ash
Liquid Effluents: None
Added Gaseous Emissions: None
Inlet SO_2 Levels: 700 to 2,500 ppm (Tested)
SO_2 Removal: 45-90+% (Tested) – Highly dependent upon operating variables
NO_x Removal: Some recent process R&D efforts include NO_x control (up to ~60%)
Particulate Removal: Integrated control capability

Commercial Application
Number/Types: None
Locations: NA
First Deployment: NA
Current Status: Active

Process Description
Lime is injected into the duct upstream of a particulate collector (fabric filter or ESP). SO_2 removal is effected during contact in both the ducting and the particulate collector. A number of processes are under development. They vary in the type of lime used (high magnesium lime vs. calcitic lime); additives (principally sodium alkali with calcitic lime); and the manner of sorbent injection (slurry or dry injection with separate partial humidification of the gas). Almost all utilize recycle of a portion of the solids collected in order to increase alkali utilization. SO_2 removal is highly mass transfer limited. Therefore, many variables impact performance: SO_2 concentration (lower is easier); type of lime (magnesium helps); degree of gas humidification (higher humidification helps); sorbent particle size (extensive grinding helps); additives (sodium alkali helps); duct contact time (longer helps); lime feed stoichiometry (higher is better); and type of particulate collector (a fabric filter is usually better).

Advantages/Disadvantages

Advantages
- Integrated SO_2 and particulate control
- Relatively low maintenance and operating labor
- Stack gas requires no reheat

Disadvantages/Limitations
- Produces a waste solid of little reuse potential
- Limitations on SO_2 removal at high inlet SO_2
- Generally high lime requirements

Principal Suppliers

Slurry Injection Systems

Supplier/Developer	Process	Absorbent
ABB/EPA	ADVACATE	"Activated" Lime (Lime/Ash)
Bechtel	CZD	High-Magnesium Lime
Dravo	HALT	Dolomitic Lime
Marsulex (GEESI)	IDS	Lime
MHI	LILAC	"Activated" Lime (Lime/Ash)

Dry Injection Systems with Humidification

Supplier/Developer	Process	Absorbent
B&W/Consol	Coolside	Lime with Sodium Alkali
EPRI	HYPAS	Lime with Sodium Alkali

Waste Producing Processes

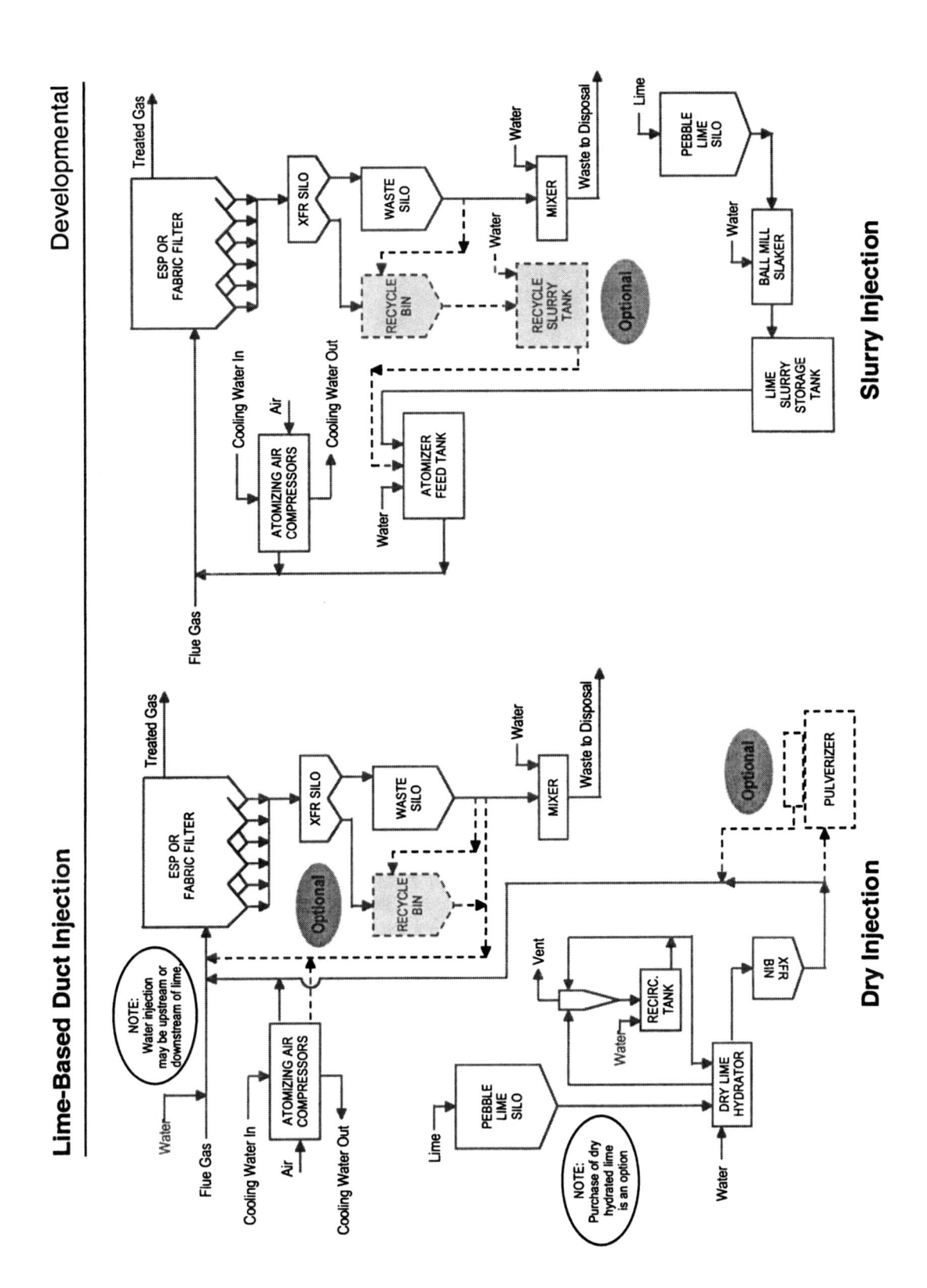

Lime-Based Economizer Injection – SO$_x$-NO$_x$-Rox-Box (SNRB)

Developmental

Hot flue gas (800°-1000°F) is contacted in a high temperature baghouse with upstream injection of alkali and ammonia for SO$_2$, NO$_X$ and particulate removal using a fabric filter with catalyst impregnated filter bags.

Process Characteristics
SO$_2$ Sorbent: Lime (with sodium alkali salts)
Principal Raw Materials: Lime (with sodium alkali salts); Ammonia (if NO$_x$ control included)
Potentially Saleable Byproducts: None
Solid Wastes: Spent sorbent and ash
Liquid Effluents: None
Added Gaseous Emissions: None
Inlet SO$_2$ Levels: 2,000 to 3,000 ppm (Tested)
SO$_2$ Removal: 50-98% (Tested)
NO$_x$ Removal: 80-95% (Tested)
Particulate Removal: Integrated control capability

Commercial Application
Number/Types: None
Locations: NA
First Deployment: NA
Current Status: Active (Pilot plant testing completed)

Process Description
The SO$_x$-NO$_x$-Rox-Box (SNRB) is an integrated SO$_x$/NO$_x$/particulate control duct dry injection process. The technology centerpiece is a high temperature fabric filter. On a coal-fired boiler it would be between an economizer and air preheater. The fabric filter is of ceramic fiber woven fabric impregnated with an SCR catalyst on the downstream (clean side) of the bag. A dry sorbent for SO$_2$ removal (lime or soda alkali) and ammonia for NO$_X$ removal are injected upstream of the baghouse. SO$_2$ is removed partly in the ductwork upstream of the baghouse, but mostly in the cake formed on the filters. The degree of SO$_2$ removal is primarily a function of the type of alkali used (sodium salts such as nacholite being more reactive than lime), operating temperature and the Ca/SO$_2$ or Na/SO$_2$ stoichiometry employed. In general, relatively high lime stoichiometry, 1.8-2.2 Ca/SO$_2$, is required to achieve "respectable" removal efficiencies (in the range of 75-90%). Efficiencies in excess of 95% require Ca/SO$_2$ stoichiometries over 2.5 which can significantly impact raw materials costs. With nacholite (natural sodium bicarbonate) roughly 90% removal can be achieved at a Na$_2$/SO$_2$ stoichiometry of about 1.0. Similarly, NO$_x$ removal is primarily a function of operating temperature and ammonia stoichiometry. The waste solids formed are removed by routine cleaning cycles and discharged through hoppers to a pneumatic conveyance system to a waste storage silo.

Advantages/Disadvantages

Advantages
- Combined SO$_X$, NO$_X$ and particulate removal
- Customization flexibility in combining gas temperature, type and load of SO$_2$ removal alkali, and ammonia feed in matching site-specific requirements
- Low operating labor potential
- A "one box" control system

Disadvantages/Limitations
- Uncertainty in long-term performance of the impregnated fabric
- Need to handle ammonia, a highly regulated hazardous chemical

Principal Developers
B&W (SNRB)   Research Cottrell (Economizer InjectionI)

Waste Producing Processes

SNRB (SO$_x$-NO$_x$-Rox-Box) — Lime-Based Economizer Injection with NO$_x$ Control   Developmental

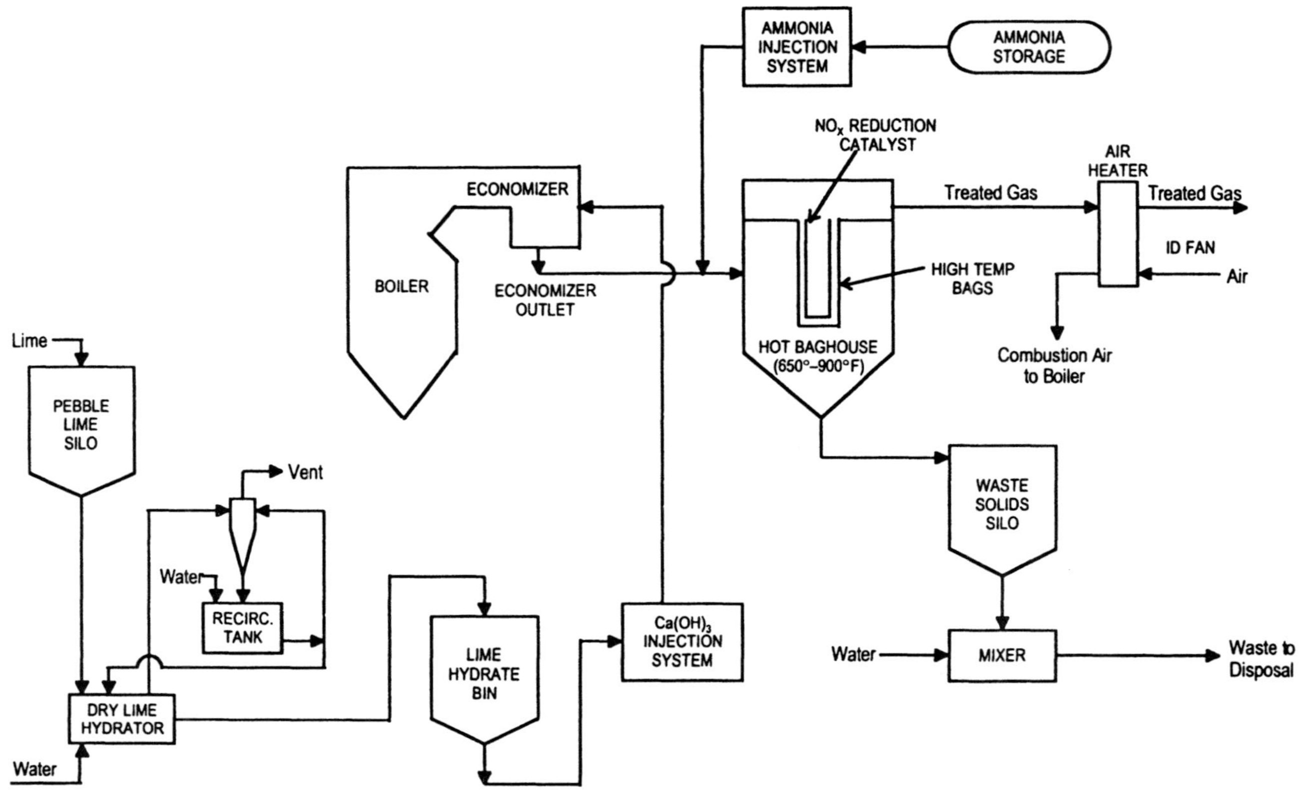

Condensing Heat Exchanger

Developmental
Flue gas is cooled in two stages of heat exchangers which, with additives, effect removal of SO_2, SO_3 and fine particulate that are then removed via a purge stream.

Process Characteristics
SO_2 Sorbent: Sodium alkali
Principal Raw Materials: Sodium alkali (e.g., soda ash); (Lime for purge treatment)
Potentially Saleable Byproducts: None
Solid Wastes: Ash and other solids from treatment of blowdown
Liquid Effluents: Purge from blowdown of heat exchangers
Added Gaseous Emissions: None
Inlet SO_2 Levels: up to ~2,000 ppm (Tested)
SO_2 Removal: up to 98% (Tested)
NO_x Removal: Nil
Particulate Removal: Particulate is removed but not as a primary particulate control device

Commercial Application
Number/Types: None
Locations: NA
First Deployment: NA
Current Status: Active (Full scale demonstration testing completed)

Process Description
The system consists of four sections: a first stage heat exchanger; a transition zone; a second stage heat exchanger; and a mist eliminator. Warm flue gas at ~300 F (e.g., from the air preheater of a conventional fossil fuel-fired combustion boiler) is fed to the first stage exchanger where most of the sensible heat is removed. The inter-stage transition section has a water or alkaline spray to saturate the gas prior to the second exchanger section and to enhance removal of acidic gases (SO_2, SO_3, and HCl) as well as fine particulate. Because of high corrosion potential in this section, it is usually constructed of fiberglass. The gas then passes to the second exchanger which operates in a condensing mode, removing latent heat from the gas. Teflon®-coated tubes are used both for corrosion protection and because Teflon® is hydrophobic. This causes condensed water to form droplets on the tubes rather than a film resulting in a "rain-out" phenomena in which the gas is flowing upward through a cascade of water droplets that remove pollutants. An alkali spray can also be used in the second section. Condensed water, demister wash and spent alkali solution are collected at the bottom of this section and discharged to a purge treatment system. Purge treatment is dependent upon the application including the amount of acidic gases collected, the amount of particulate removed, the availability and cost of neutralization alkali and local constraints on wastewater discharge.

Advantages/Disadvantages

Advantages
- Combine SO_2 and SO_3 removal
- Integrated SO_x and low level particulate control
- Relatively low maintenance and operating labor
- Little to no scaling or plugging potential
- Heat recovery improving plant thermal efficiency
- Lower plume visibility due to lower water content of gas

Disadvantages/Limitations
- Produces a purge liquor requiring treatment
- Not applicable to high SO_2 levels for high removal efficiencies

Principal Developers
B&W/CHX (IFGT)

Waste Producing Processes

Condensing Heat Exchanger Process

Developmental

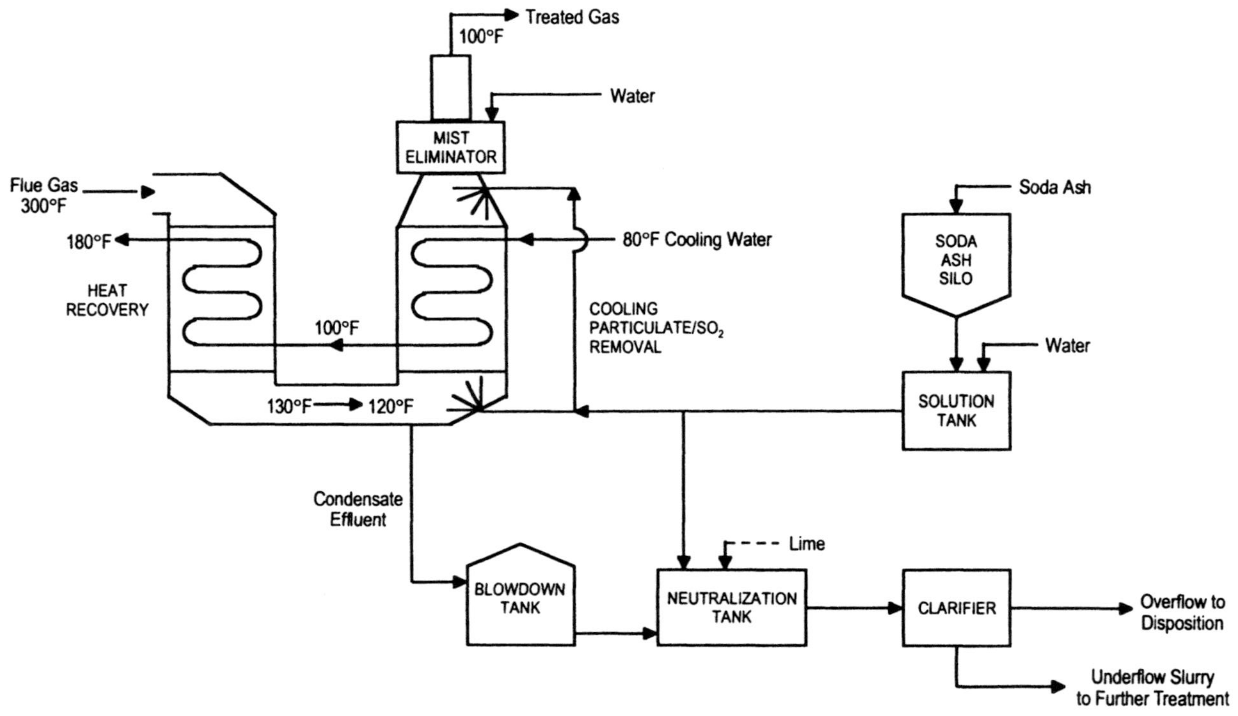

SECTION 3

FGD TECHNOLOGY PROFILES
Byproduct Processes

Lime Slurry Forced Oxidation

Commercial
Absorption of SO$_2$ in lime slurry with air oxidation of the precipitated solids followed by gypsum byproduct solids separation and dewatering.

Process Characteristics
SO$_2$ Sorbent: A mixture of lime and calcium sulfite/sulfate (sometimes with additives such as magnesium salts if Mg-promoted lime is used)
Principal Raw Materials: Lime (sometimes high magnesium content such as Dravo's Thiosorbic, lime); Additives in some applications (e.g., formic acid)
Potentially Saleable Byproducts: Gypsum
Solid Wastes: None
Liquid Effluents: Purge liquor usually required for commercial grade gypsum
Other Gaseous Emissions: None
Inlet SO$_2$ Levels: 1,500 to 3,000 ppm (Commercial Operations)
SO$_2$ Removal: Typical Design Range: 90-95%; Maximum at Highest Inlet SO$_2$: ~96%
NO$_x$ Removal: Nil (Intentional oxidation precludes most additives)
Particulate Removal: Integrated control capability unless producing commercial grade gypsum

Commercial Application
Number/Types: Several: Utility and industrial boilers; process furnaces
Locations: US; Europe
First Deployment: Mid-1980s
Current Status: Active

Process Description
The first significant commercialization of this technology was by Saarberg-Holter on a utility boiler in Germany. The approach utilized formic acid as a buffer for SO$_2$ removal and integrated aeration of the slurry. Other systems have evolved primarily on the industrial scale and in Europe mostly without additives. Forced oxidation in lime slurry systems has only recently taken hold in utility boilers in the US, generally as retrofits to conventional lime scrubbing in order to produce gypsum rather than waste calcium sulfite/sulfate solids which have no market value and can be difficult to dispose. In a typical system, the gas contactor consists of a multi-level spray tower or deluge type device (e.g., venturi scrubber) or combination of both usually with an internal recirculation tank at the bottom. Lime is fed to the recirculation tank and slurry from the tank is recirculated through the contactor at a high L/G ratio. Two approaches are now commonly employed in the US to effect oxidation of the solids. One involves removal of the sulfite-rich slurry from the scrubber and oxidation with air prior to neutralization with lime. This allows a relatively low pH for continued dissolution and oxidation of the precipitated calcium sulfite. Where pH adjustment is required, sulfuric acid is used. The solids are then separated from the liquor by hydroclones or conventional thickening followed by filtration. This approach is geared toward producing saleable gypsum – the majority for wall board; the fines for cement manufacture. The other involves oxidation of partially neutralized solids from the scrubber circuit usually in a separate aeration tank. The solids are then usually pumped to a disposal impoundment. The separated liquor may or may not be recycled to the scrubbing circuit.

Advantages/Disadvantages

Advantages
- Retrofittable to conventional lime scrubbers producing waste solids
- Can produce commercial grade gypsum
- Significant design advances minimize scaling and plugging problems
- Capable of integrated SO$_2$ and particulate control

Disadvantages/Limitations
- Commercial grade gypsum usually requires a purge liquor treatment system
- Achieving high removal efficiencies in high sulfur applications requires close chemistry control and possible use of additives
- Gypsum quality equivalent to limestone-based technology not demonstrated

Principal Suppliers
Dravo   Radian

Byproduct Processes

Lime Slurry Forced Oxidation

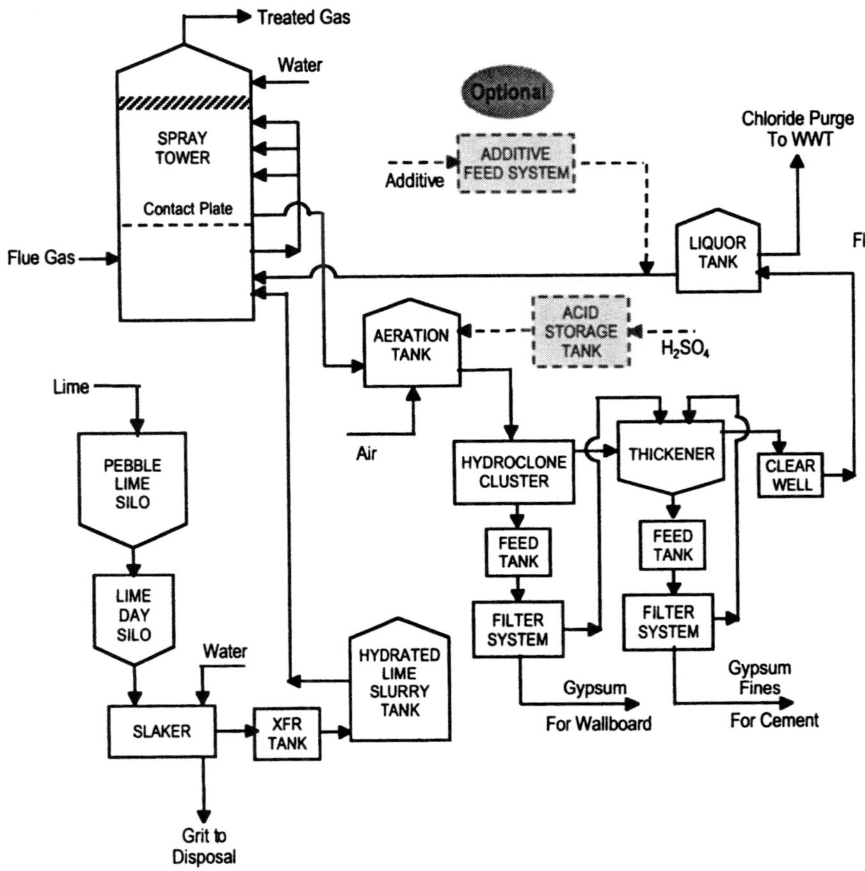

Version 1: Commercial Gypsum

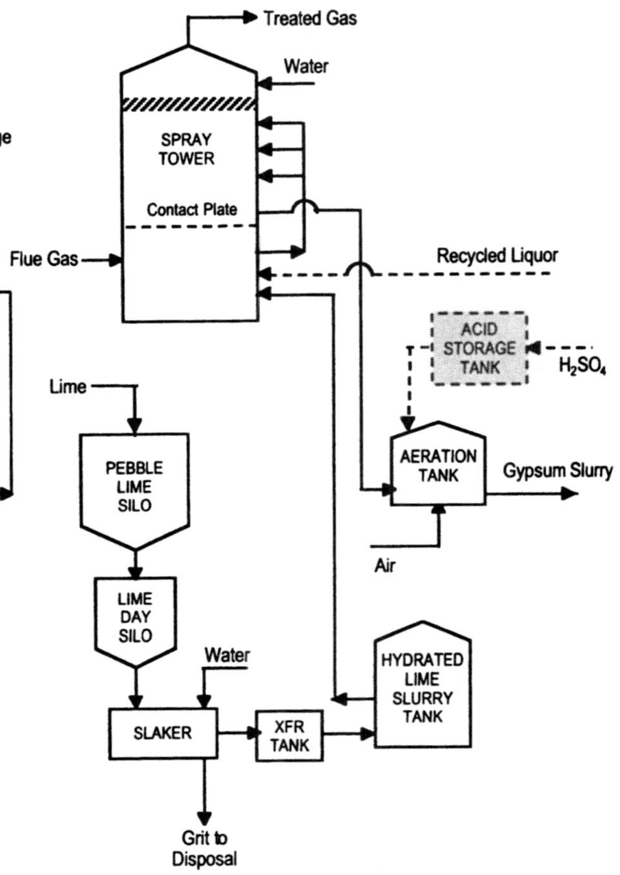

Version 2: Gypsum Slurry Disposal Commercial

Limestone Slurry Forced Oxidation

Commercial
Absorption of SO$_2$ in a limestone slurry, air oxidation of precipitated solids to gypsum, then solids separation and washing to produce commercial grade gypsum.

Process Characteristics
SO$_2$ Sorbent: A mixture of limestone, calcium sulfite and gypsum
Principal Raw Materials: Limestone
Potentially Saleable Byproducts: Gypsum
Solid Wastes: None
Liquid Effluents: Purge liquor usually required for commercial grade gypsum
Other Gaseous Emissions: None
Inlet SO$_2$ Levels: <1,000 to ~4,500 ppm (Commercial Operations)
SO$_2$ Removal: Typical Design Range: 80-98%; Maximum at Highest Inlet SO$_2$: ~95%
NO$_x$ Removal: Nil (Intentional oxidation precludes most additives)
Particulate Removal: Integrated control capability unless producing commercial grade gypsum

Commercial Application
Number/Types: Many: Utility & industrial boilers; refineries; metallurgical plants
Locations: North America; Europe; Far East
First Deployment: 1980s (Current generation of technology)
Current Status: Active

Process Description
In a typical system, the gas contactor consists of a multi-level spray tower with an internal recirculation tank at the bottom. Limestone is fed to the recirculation tank and slurry from the tank is recirculated through the sprays at a high L/G ratio. Air is also sparged into the bottom of the tank to effect conversion of absorbed SO$_2$ to SO$_3$ and precipitation of gypsum. In current designs, gypsum is continually removed from the scrubbing circuit by circulating scrubber slurry through hydroclones. This allows removal of the largest crystals with return of fines to the scrubber. The gypsum is usually either centrifuged or filtered, although in some cases it is sent directly to onsite stack-out piles. Recovered liquor is returned to the scrubber and used for preparation of limestone slurry. There are numerous variations in chemistry and configuration. Notable among these are the Chiyoda 121 Process that utilizes a jet bubbling reactor for gas contacting rather than a spray tower and the Noell/RC technology that utilizes two stage scrubbing. Some systems use additives for buffering to enhance SO$_2$ removal. Production of commercial grade gypsum usually requires washing of the gypsum as well as purging of liquor to control buildup of impurities that would contaminate the product.

Advantages/Disadvantages

Advantages
- Uses low cost limestone as the reagent
- Can produce commercial grade gypsum
- Significant design advances now minimize scaling and plugging problems
- Capable of integrated SO$_2$ and particulate control

Disadvantages/Limitations
- Generally limited to ~4,000 ppm inlet SO$_2$ although designs are proposed for higher levels
- Removal efficiencies limited to ~97% at highest inlet SO$_2$ levels
- Commercial grade gypsum usually requires a purge liquor treatment system

Principal Suppliers

ABB	Kawasaki	Mitsui	Radian (Retrofits)
B&W	Lentjes	Nippon Kokan	Riley
Chiyoda (121 Process)	Marsulex	Noell/RC	Saarberg-Holter
Deutsche Babcock	Mitsubishi/PureAir	Procedair	Thyssen

Byproduct Processes

Limestone Slurry Forced Oxidation — Commercial

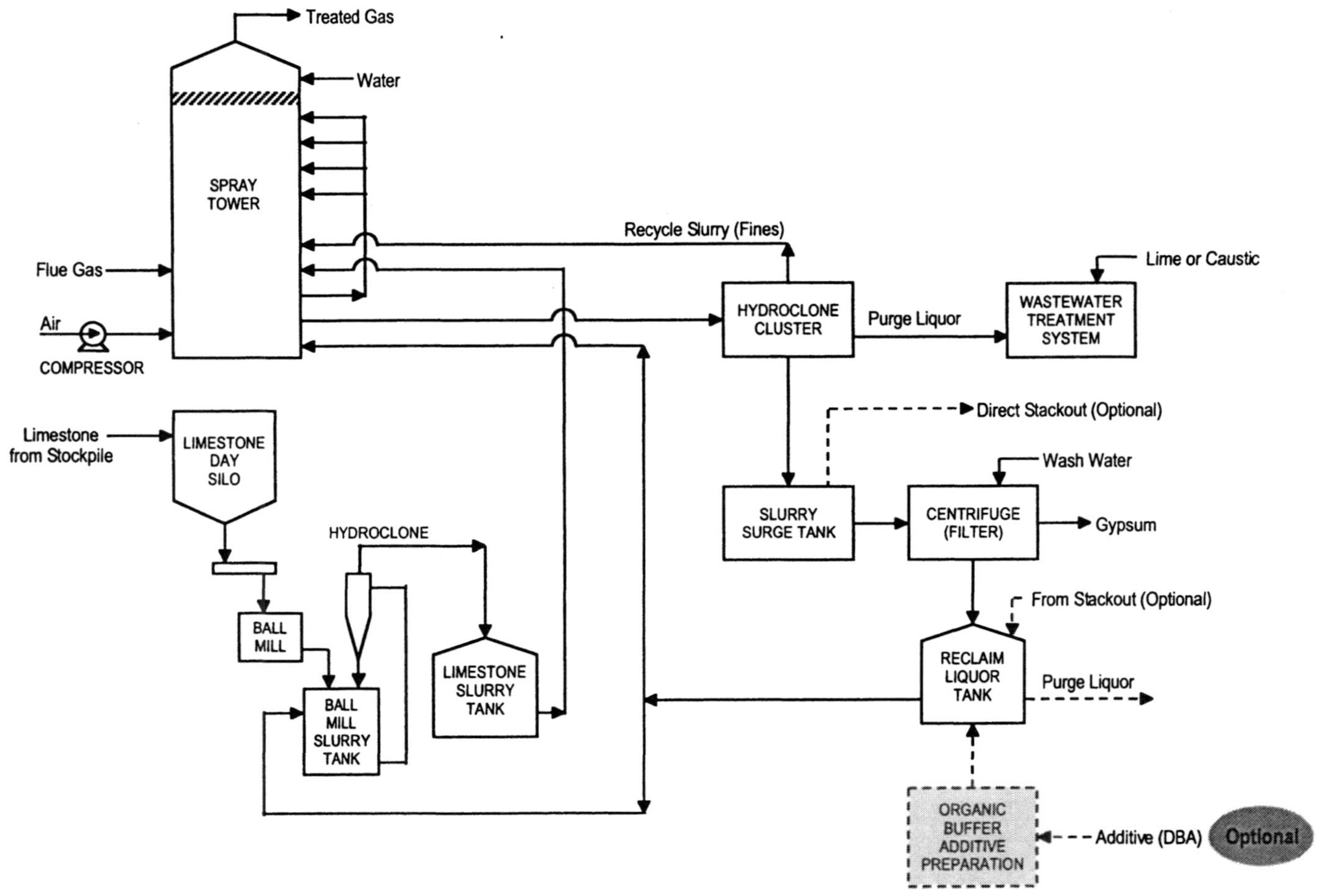

Dilute Sulfuric Acid to Gypsum

Commercial
SO₂ absorption in dilute sulfuric acid followed by oxidation of the absorbed SO₂ using air and then reaction of the acid with limestone to produce gypsum.

Process Characteristics
SO₂ Sorbent: Dilute sulfuric acid
Principal Raw Materials: Soluble catalyst makeup (e.g., iron sulfate); Lime or soda ash for treating purge liquor, if required
Potentially Saleable Byproducts: Gypsum
Solid Wastes: A small amount from treatment of purge liquor, if required
Liquid Effluents: Treated purge liquor
Added Gaseous Emissions: None
Inlet SO₂ Levels: <1,000 to 2,000 ppm (Commercial Operations)
SO₂ Removal: Typical Design Range: 85-90%
Maximum at Highest Inlet SO₂: ~95%
NOₓ Removal: Nil
Particulate Removal: Upstream decoupled particulate removal required

Commercial Application
Number/Types: Many: Over a dozen oil-fired utility & industrial boilers
Locations: Far East (Japan)
First Deployment: 1970s
Current Status: Generally inactive although plants still in operation

Process Description
The Chiyoda CT-101 and 102 technologies were the most widely deployed with 14 installations on oil-fired boilers in Japan by 1980. Both processes use a combined absorption and oxidation tower with an internal recirculation tank. Flue gas is contacted in the annular section of the tower with a dilute sulfuric acid solution (2-3%) which absorbs SO₂. Acid solution is collected in the bottom where air injection carries a spent acid through a central tray column where SO₂ is oxidized to SO₃. A portion of the recirculating acid is continuously withdrawn to a crystallizer where it is reacted with limestone to produce gypsum. The gypsum is removed in a centrifuge. The recovered is liquor is first clarified to remove fines that are recirculated to the crystallizer and then returned to the scrubber circuit.

Advantages/Disadvantages

Advantages
- Essentially eliminates scaling and plugging problems in scrubber
- Produces commercial grade gypsum

Disadvantages/Limitations
- Generally low SO₂ removal efficiency capability by today's standards
- Highly corrosive solutions require alloy materials which greatly increase the capital cost compared to direct limestone scrubbing systems
- Purge liquor requires treament

Principal Suppliers
Chiyoda (101 and 102 Processes)   Showa Denko

Byproduct Processes

Chiyoda 101 Process — Dilute Acid to Gypsum

Commercial

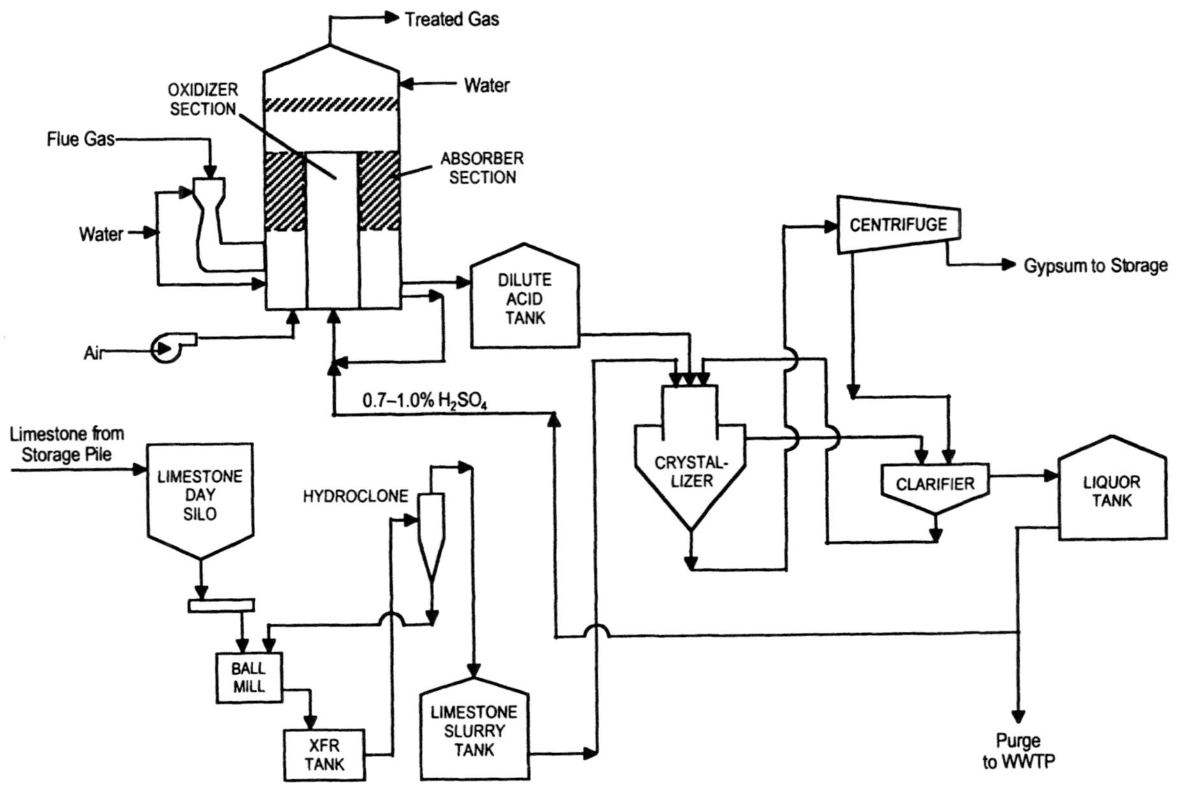

# Sodium/Lime Dual Alkali (Dilute Mode)

Commercial
Absorption of SO_2 in a sodium sulfite/bisulfite solution, oxidation of the spent solution, then reaction with lime to regenerate the absorbent and produce gypsum. Softening of the regenerated solution may also be required.

Process Characteristics
SO_2 Sorbent: Sodium hydroxide solution (with sodium sulfate)
Principal Raw Materials: Lime (hydrated or quick lime); Caustic or soda ash; Carbon dioxide (for softening, if required)
Potentially Saleable Byproducts: Gypsum; ($CaCO_3$ if softening is used)
Solid Wastes: None
Liquid Effluents: None
Added Gaseous Emissions: None
Inlet SO_2 Levels: <100 to 12,000 ppm (Commercial Operations)
SO_2 Removal: Typical Design Range: 95- 99.5%; Maximum Achievable at Highest SO_2: 99.9+%
NO_x Removal: Nil
Particulate Removal: Integrated control capability unless producing commercial grade gypsum

Commercial Application
Number/Types: Two ore smelters; one industrial boiler (inactive)
Locations: US
First Deployment: 1975
Current Status: Active

Process Description
Flue gas is contacted usually in a "deluge" type scrubber (e.g., venturi) with a solution of sodium hydroxide which absorbs the SO_2. Spent solution of sodium sulfite/bisulfite is continuously withdrawn to an oxidizer where the solution is neutralized with recycled absorbent solution and the sulfite oxidized to sodium sulfate. The sulfate solution is then sent to a reactor where it is contacted with hydrated lime, precipitating gypsum and regenerating sodium hydroxide. Gypsum is removed by thickening and filtration, and the regenerated solution is returned to the scrubber. Depending upon the flue gas composition and type of scrubber used, it may be necessary to "soften" the regenerated solution prior to return to the scrubber due to the relatively high dissolved calcium levels after regeneration (1500-2000 ppm as $CaCO_3$). This is accomplished in a separate softening loop using makeup soda ash and carbon dioxide. The gypsum (and $CaCO_3$) solids are usually washed to recover sodium salts. Sodium makeup varies from <1% to ~3% of the absorbed SO_2 depending upon the extent of wash.

Advantages/Disadvantages

Advantages
- Clear solution scrubbing minimizes scaling and plugging problems
- High removal efficiencies for a wide range of inlet SO_2 levels
- Tolerant of wide fluctuations in inlet SO_2
- Tolerant of wide ranges of absorbent oxidation up to 100%
- Capable of integrated SO_2 and particulate control

Disadvantages/Limitations
- High liquor flow rates due to the dilute solution increases capital costs for the regeneration circuit
- Use of solution softening increases lime consumption by 20-40% and increases the capital and operating costs
- Prone to scaling and plugging problems in the regeneration circuit

Principal Suppliers
Arthur D. Little

Byproduct Processes

Sodium/Lime Dual Alkali (Dilute Mode) — Commercial

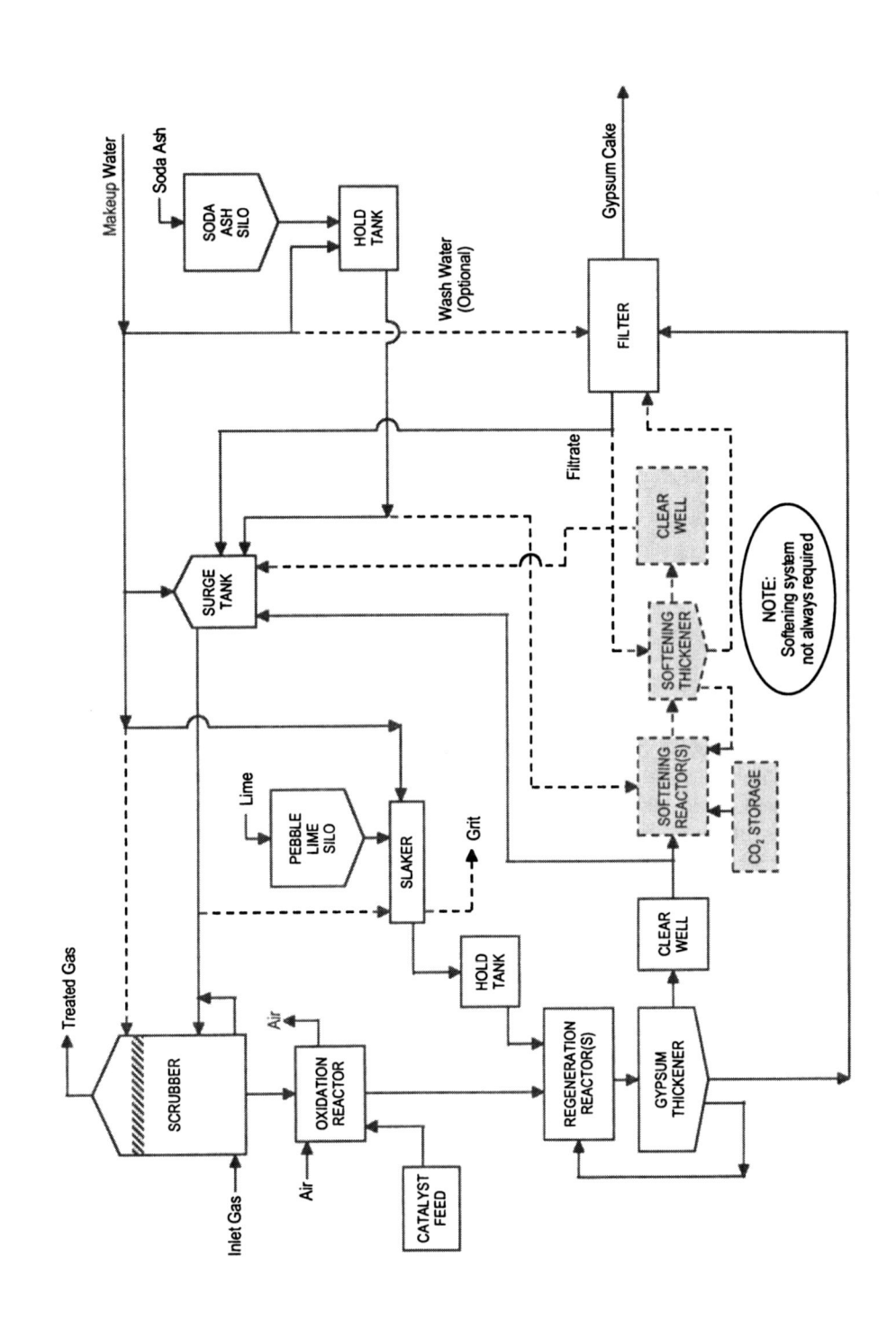

NOTE: Softening system not always required

Sodium/Limestone Dual Alkali with H_2SO_4 Conversion

Commercial
Absorption of SO_2 in a sodium sulfite/bisulfite solution, reaction of spent solution with limestone to regenerate the absorbent and precipitate calcium sulfite which is then oxidized under acid conditions to produce gypsum.

Process Characteristics
SO_2 Sorbent: Sodium sulfite/bisulfite solution (with sodium sulfate)
Principal Raw Materials: Limestone; Caustic or soda ash; Sulfuric acid
Potentially Saleable Byproducts: Gypsum
Solid Wastes: None
Liquid Effluents: None
Inlet SO_2 Levels: 1,000 to 3,000 ppm (Commercial Operations)
Added Gaseous Emissions: None
SO_2 Removal: Typical Design Range: 90-95%; Maximum Achievable at Highest SO_2: 99+%
NO_x Removal: Nil
Particulate Removal: Upstream decoupled particulate removal required

Commercial Application
Number/Types: Several: Utility & industrial boilers
Locations: Far East (Japan)
First Deployment: 1970s
Current Status: Unknown

Process Description
Flue gas is contacted in an absorber with a solution of sodium sulfite/bisulfite/bicarbonate which absorbs the SO_2. Spent solution of sodium sulfite/bisulfite is continuously withdrawn to a reactor system where it is contacted with limestone. The limestone precipitates calcium sulfite hemihydrate (containing a small amount of calcium sulfate in solid solution) and regenerates the absorbent solution. The precipitated solids are removed by thickening and the regenerated solution is returned to the scrubber. A portion of the thickened slurry is fed to a converters/reactor where it is reacted with sulfuric acid to remove sulfate by precipitation of gypsum. The rest of the thickener underflow is filtered. The filter cake is then oxidized also to produce gypsum. Both reactor effluents are then dewatered by centrifugation or filtration to produce commercial grade gypsum.

Advantages/Disadvantages

Advantages
- Clear solution scrubbing minimizes scaling and plugging problems
- High removal efficiencies for a wide range of inlet SO_2 levels
- Tolerant of wide fluctuations in inlet SO_2
- Produces commercial grade gypsum
- Tolerant of a wide range of absorbent oxidation
- Uses low cost limestone

Disadvantages/Limitations
- Relatively complicated process configuration of moderately high cost
- Requires the use of sulfuric acid which increases the consumption of limestone.

Principal Suppliers
Kureha-Kawasaki   Showa Denko-Ebara   Tsukishima

Sodium/Limestone Dual Alkali with Sulfuric Acid Conversion   Commercial

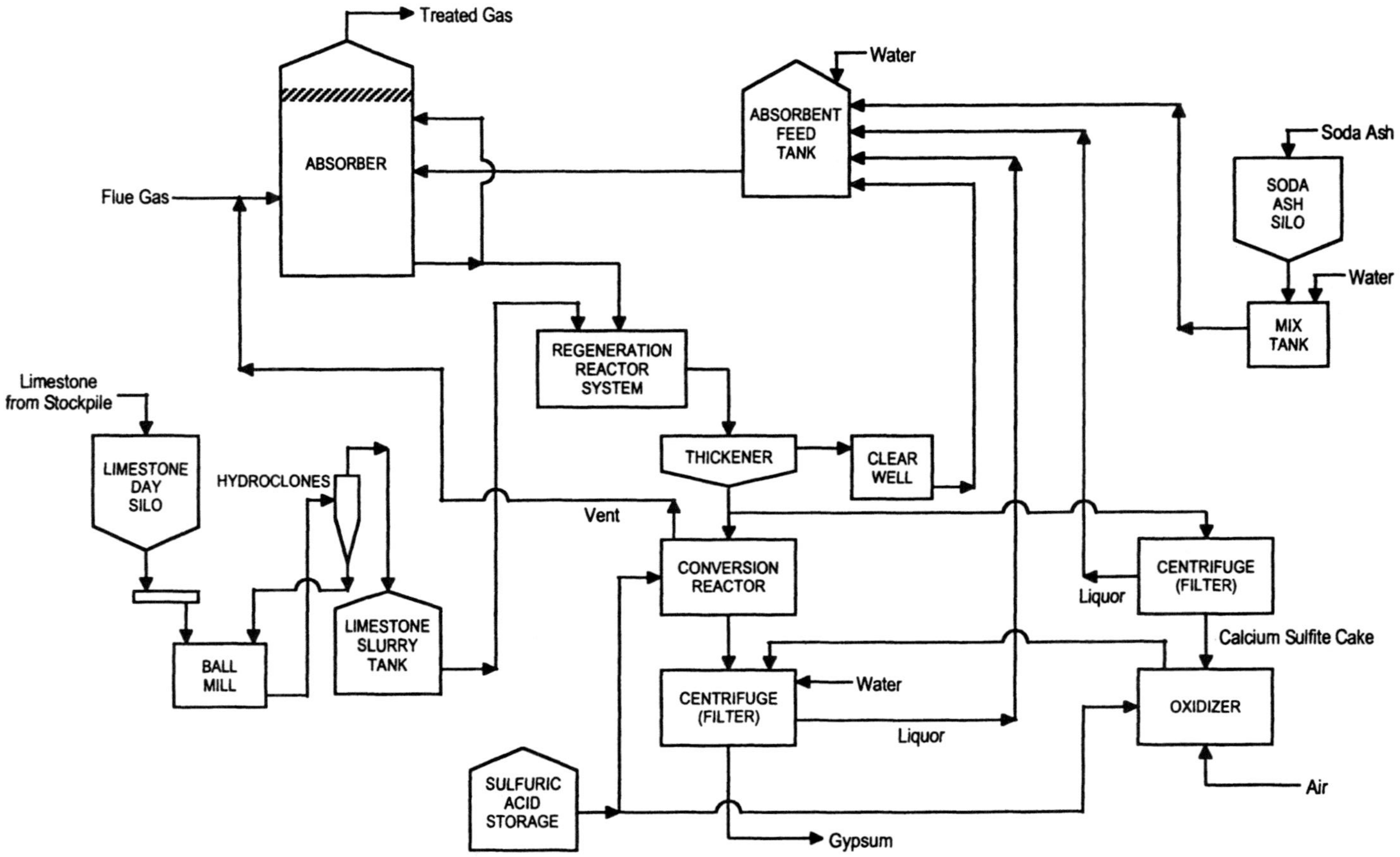

Dowa Process – Aluminum Sulfate/Limestone Dual Alkali

Commercial
Absorption of SO$_2$ in aluminum sulfate solution, oxidation of the spent solution, then reaction with limestone to regenerate the absorbent and precipitate gypsum.

Process Characteristics
SO$_2$ Sorbent: Basic aluminum sulfate solution
Principal Raw Materials: Limestone; Aluminum sulfate
Potentially Saleable Byproducts: Gypsum
Solid Wastes: None
Liquid Effluents: Purge liquor may be required to control impurities for commercial gypsum
Added Gaseous Emissions: Potential for some SO$_2$/SO$_3$ release from the oxidation tower if not vented to the absorber
Inlet SO$_2$ Levels: <1,000 to 25,000 ppm (Commercial Operations)
SO$_2$ Removal: Typical Design Range: 85-98%; Maximum Achievable at Highest SO$_2$: 98%
NO$_x$ Removal: Nil
Particulate Removal: Upstream decoupled particulate removal required

Commercial Application
Number/Types: Many: Oil-fired boilers; ferrous and nonferrous smelters/roasters; acid plants
Locations: Far East (Japan)
First Deployment: Early 1970s
Current Status: Active

Process Description
Flue gas is contacted in a packed tower absorber with a solution of basic aluminum sulfate which absorbs the SO$_2$. Spent scrubber solution is continuously withdrawn to an oxidation tower where it is contacted with air to convert the sulfite to sulfate. Most of the oxidized liquor is recycled to the absorber to provide the high L/G required for efficient SO$_2$ removal. A slipstream of oxidized liquor is sent to a neutralization/recovery loop where it is first used to dissolve precipitated aluminum hydroxide. It is then neutralized with limestone to regenerate the basic aluminum sulfate solution and precipitate gypsum. Gypsum is removed by conventional thickening and filtration. A bleed stream of solution is also withdrawn to control buildup of chlorides and other contaminants.

Advantages/Disadvantages

Advantages
- Clear solution scrubbing minimizes scaling and plugging problems
- High removal efficiencies for a wide range of inlet SO$_2$ levels
- Tolerant of wide fluctuations in inlet SO$_2$
- Produces commercial grade gypsum
- Tolerant of a wide range of absorbent oxidation

Disadvantages/Limitations
- Relatively complicated process configuration of moderately high cost

Principal Suppliers
Dowa Mining Co.

Byproduct Processes 63

Dowa Process—Aluminum Sulfate/Limestone Dual Alkali Commercial

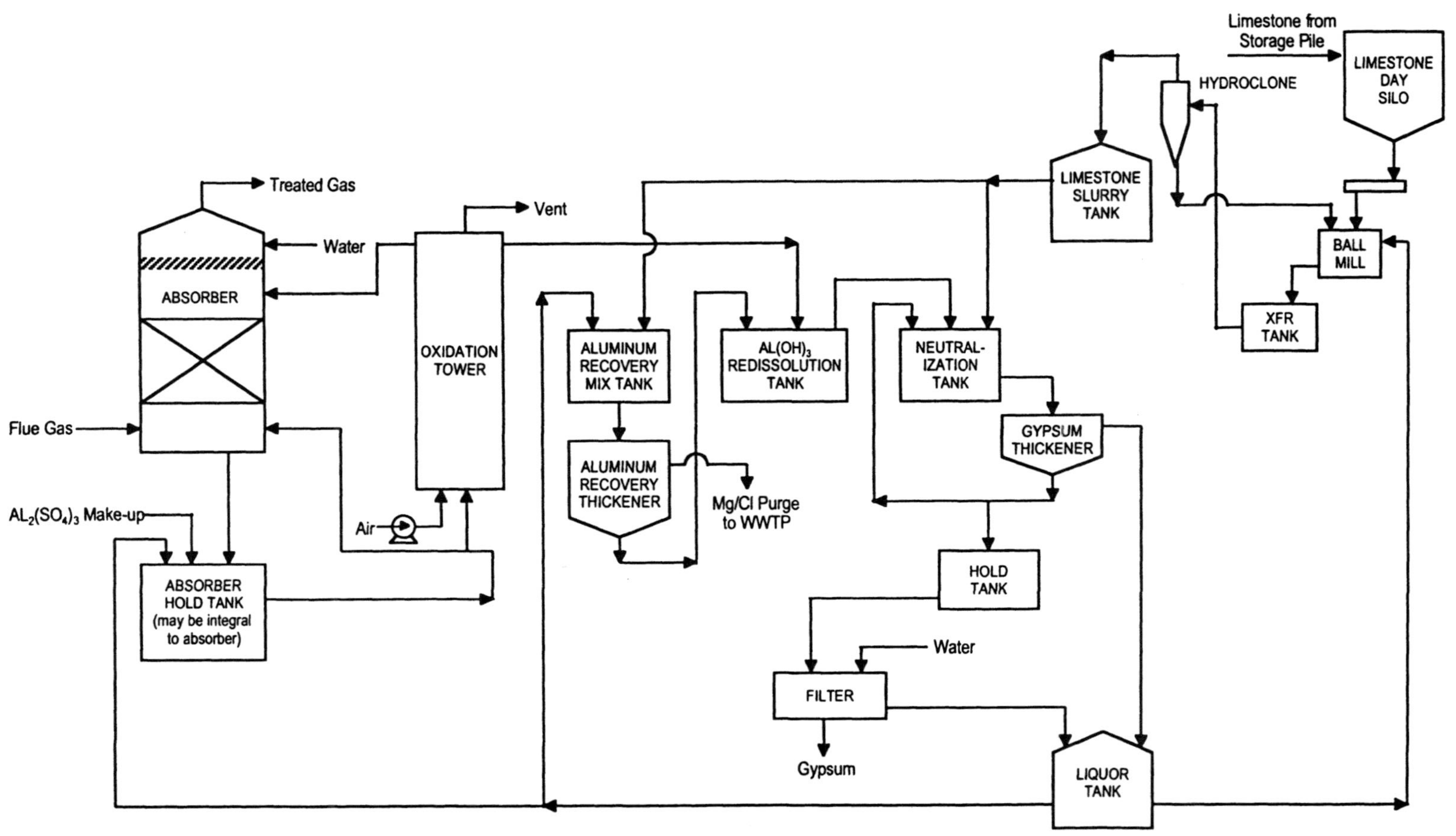

Kurabo Process – Ammonia/Lime Dual Alkali

Commercial
Absorption of SO$_2$ in an ammonium salt solution, oxidation of the absorbent solution, then reaction with lime to regenerate the absorbent and precipitate gypsum.

Process Characteristics
SO$_2$ Sorbent: Ammonium sulfate/bisulfate/bisulfite solution
Principal Raw Materials: Lime; Ammonia
Potentially Saleable Byproducts: Gypsum
Solid Wastes: None
Liquid Effluents: Purge liquor may be required to control impurities for commercial gypsum
Added Gaseous Emissions: Potential for some SO$_2$/SO$_3$ or ammonia/ammonium sulfate release from the oxidation tower if not vented to the absorber. Potential for "blue plume" in stack gas discharge
Inlet SO$_2$ Levels: ~1,500 ppm (Commercial Operations)
SO$_2$ Removal: Typical Design Range: 85-90%; Maximum Achievable at Highest SO$_2$: 93% (Demonstrated)
NO$_x$ Removal: Nil
Particulate Removal: Integrated capability unless commercial grade gypsum produced

Commercial Application
Number/Types: Several – Industrial oil-fired boilers
Locations: Japan
First Deployment: Early 1980s
Current Status: Unknown

Process Description
Flue gas is contacted with a solution of ammonium sulfate/bisulfate/bisulfite which absorbs the SO$_2$. The solution is maintained at a pH of 3-4 by continuously recycling the scrubber solution through an air-blown oxidizer to convert the sulfite to sulfate. While this reduces the absorptive capacity of the solution necessitating high L/G ratios. However, it suppresses the vapor pressure of ammonia which avoids ammonium salt plumes characteristic of most ammonia scrubbers. Spent scrubber solution is sent to a reactor where it is contacted with hydrated lime, precipitating gypsum and regenerating ammonium hydroxide solution. Gypsum is removed by conventional thickening and filtration, and the regenerated solution is returned to the scrubber.

Advantages/Disadvantages

Advantages
- Clear solution scrubbing minimizes scaling and plugging problems
- Produces commercial grade gypsum
- Tolerant of a wide range of absorbent oxidation

Disadvantages/Limitations
- Need to handle ammonia, a highly regulated hazardous chemical
- Difficult to handle high inlet SO$_2$ levels economically

Principal Suppliers
Kurabo

Byproduct Processes

Kurabo Process — Ammonia/Lime Dual Alkali

Commercial

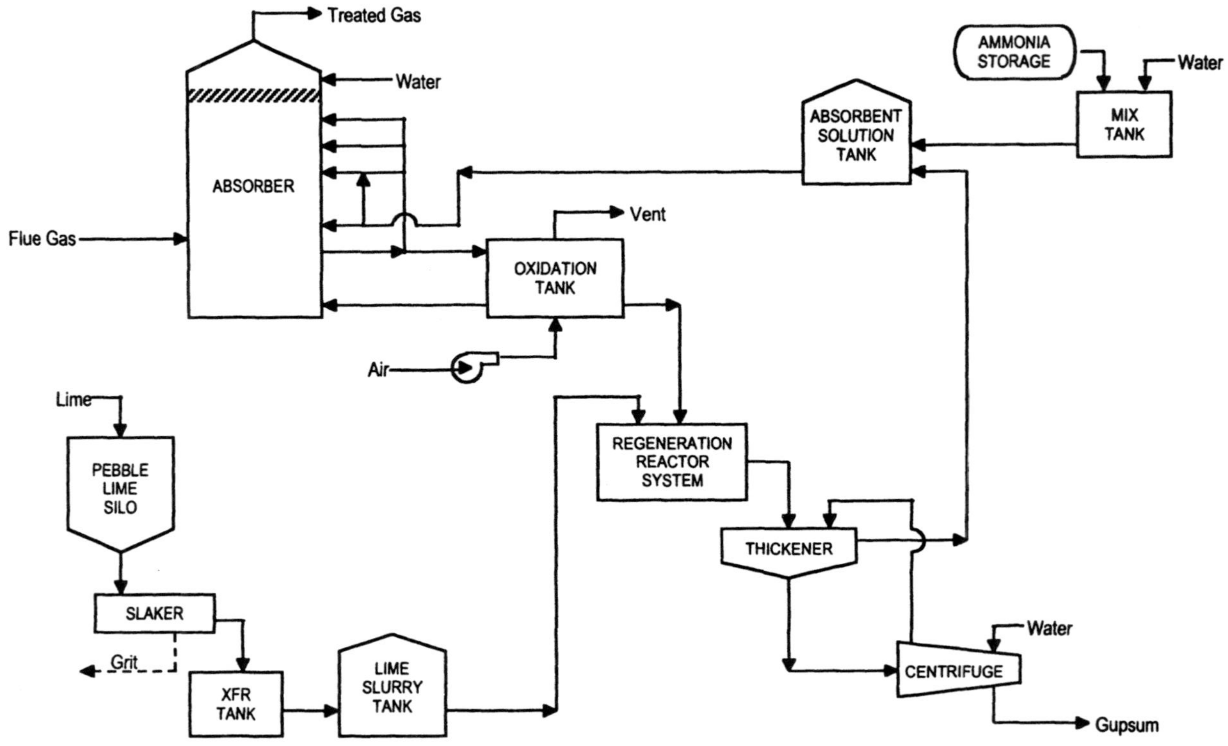

Thioclear® Process – Magnesium Solution/Lime Dual Alkali

Commercial
Absorption of SO$_2$ in magnesium sulfite solution, oxidation of the spent solution, then reaction with lime to regenerate the absorbent and precipitate gypsum.

Process Characteristics
SO$_2$ Sorbent: Magnesium sulfite/bisulfite solution
Principal Raw Materials: Lime; Magnesium hydroxide (If low Mg-lime is used)
Potentially Saleable Byproducts: Gypsum; Magnesium hydroxide (If high Mg-lime is used)
Solid Wastes: None
Liquid Effluents: Purge liquor may be required to control impurities for commercial gypsum
Added Gaseous Emissions: None
Inlet SO$_2$ Levels: ~2,000 ppm (Commercial Operations)
SO$_2$ Removal: Typical Design Range: 90-95%; Maximum Achievable at Highest SO$_2$: 97+% (Projected)
NO$_x$ Removal: Nil
Particulate Removal: Integrated capability unless commercial grade gypsum produced

Commercial Application
Number/Types: One set of industrial boilers and one utility boiler
Locations: US
First Deployment: 1998
Current Status: Active

Process Description
Flue gas is contacted in a tray or disk/doughnut tower with a solution of magnesium sulfite/bisulfite which absorbs the SO$_2$. Spent solution of magnesium sulfite/bisulfite is continuously withdrawn to an oxidizer where the solution is neutralized with recycled absorbent solution and the sulfite oxidized to sulfate. Some amount of precipitated calcium sulfate is also recycled to the oxidizer to minimize scaling from calcium sulfate supersaturation. The sulfate solution is then sent to a reactor where it is contacted with hydrated lime, precipitating gypsum and regenerating sodium hydroxide. Gypsum is removed by thickening and filtration, and the regenerated solution is returned to the scrubber. The gypsum is usually washed to recover magnesium salts. Magnesium makeup requirements vary primarily with the type of lime used. If calcitic (low-Mg) lime is used, then magnesium hydroxide addition is required equal to about 2-3% of the SO$_2$ absorbed. If high magnesium content lime is used (3-6% Mg), then no magnesium makeup is required, and magnesium hydroxide is actually produced as a byproduct value. A bleed stream of solution is also withdrawn to control buildup of chlorides and other contaminants.

Advantages/Disadvantages

Advantages
- Clear solution scrubbing minimizes scaling and plugging problems
- High removal efficiencies for a wide range of inlet SO$_2$ levels
- Produces commercial grade gypsum
- Tolerant of a wide range of absorbent oxidation

Disadvantages/Limitations
- Relatively complicated process configuration of moderately high cost
- Not as tolerant as sodium based dual alkali to wide fluctuations in inlet SO$_2$ levels
- Prone to scaling and plugging problems in the regeneration circuit

Principal Suppliers
Dravo

Byproduct Processes

Dravo Thioclear® Process — Magnesium Solution/Lime Dual Alkali   Commercial

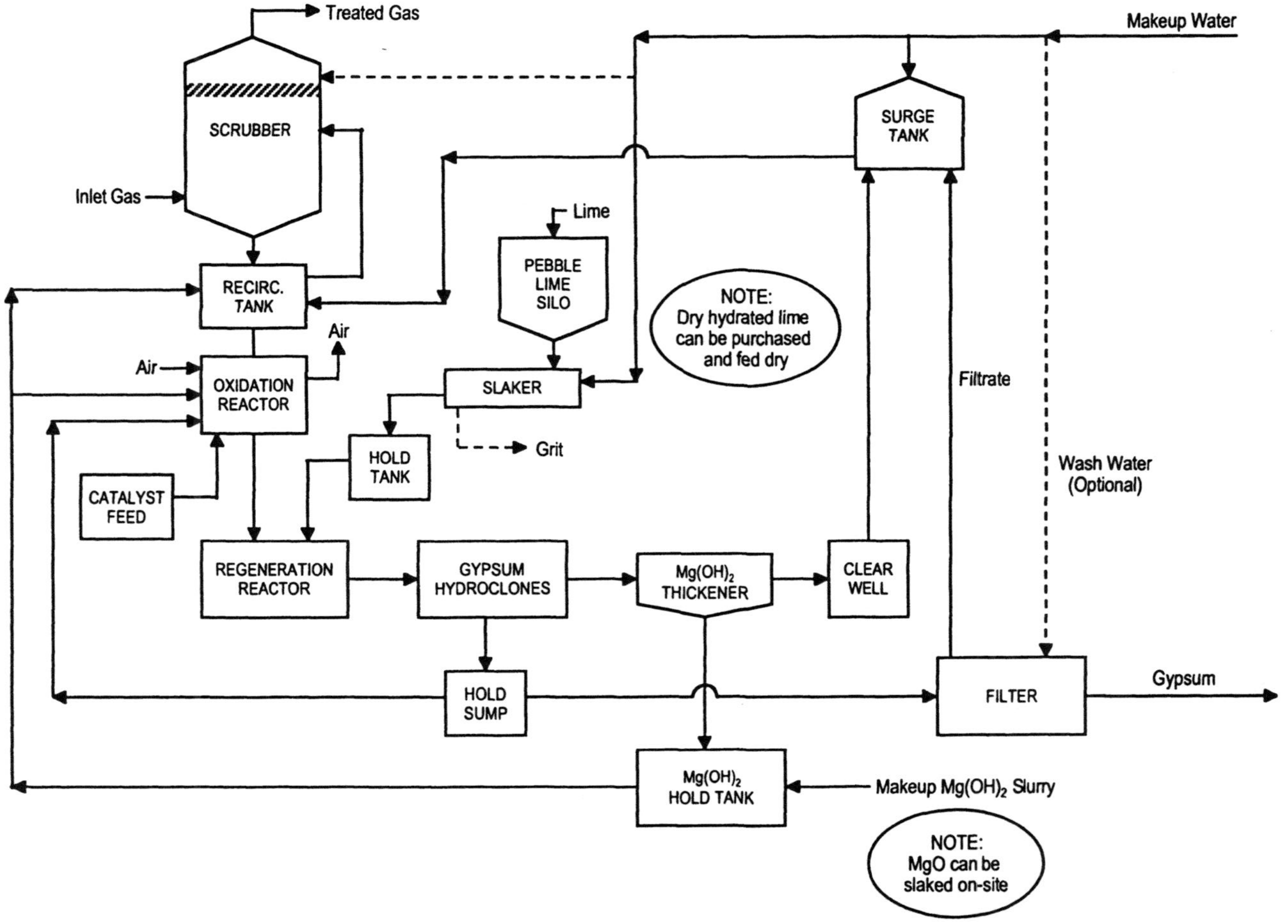

Kawasaki Process – Magnesium Slurry/Lime Dual Alkali

Commercial
Absorption of SO_2 in a slurry of magnesium sulfite, oxidation of the spent solution, then reaction with lime to regenerate the absorbent slurry and precipitate gypsum.

Process Characteristics
SO_2 Sorbent: Magnesium (and calcium) sulfite/bisulfite solution/slurry
Principal Raw Materials: Lime; Magnesium hydroxide
Potentially Saleable Byproducts: Gypsum
Solid Wastes: None
Liquid Effluents: Purge liquor may be required to control impurities for commercial gypsum
Added Gaseous Emissions: None
Inlet SO_2 Levels: 1,500 to 2,000 ppm (Commercial Operations)
SO_2 Removal: Typical Design Range: 85-95%; Maximum Achievable at Highest SO_2: 97%
NO_x Removal: Nil
Particulate Removal: Integrated capability unless commercial grade gypsum produced

Commercial Application
Number/Types: Several: Oil fired boilers
Locations: Japan
First Deployment: 1975
Current Status: Unknown

Process Description
Flue gas is contacted in a multiple spray tower with a slurry of magnesium and calcium sulfite and magnesium hydroxide which absorbs the SO_2 forming soluble magnesium and calcium bisulfite. Spent slurry is continuously withdrawn to an oxidizer where the solution is contacted with air to convert all of the sulfite and bisulfite to sulfate with the resulting precipitation of gypsum. The oxidizer is vented to the absorber due to the potential for some SO_2 is offgasing at the low pH. Gypsum is removed from the oxidized slurry by conventional thickening and centrifugation. A portion of the recovered liquor is returned to the scrubber. The remainder is sent to a regeneration tank where it is reacted with lime. This converts the $MgSO_4$ to $Mg(OH)_2$ and precipitates additional gypsum. The regenerated slurry is then fed to the scrubber.

Advantages/Disadvantages

Advantages
- Minimal scale potential in the scrubber with pre-saturated slurry and low operating pH
- Produces commercial grade gypsum
- Tolerant of a wide range of absorbent oxidation

Disadvantages/Limitations
- Not as tolerant as sodium based dual alkali to wide fluctuations in inlet SO_2 levels due to lower buffering capacity and lower operating pH

Principal Suppliers
Kawasaki

Kawasaki Process — Magnesium Slurry/Lime Dual Alkali   Commercial

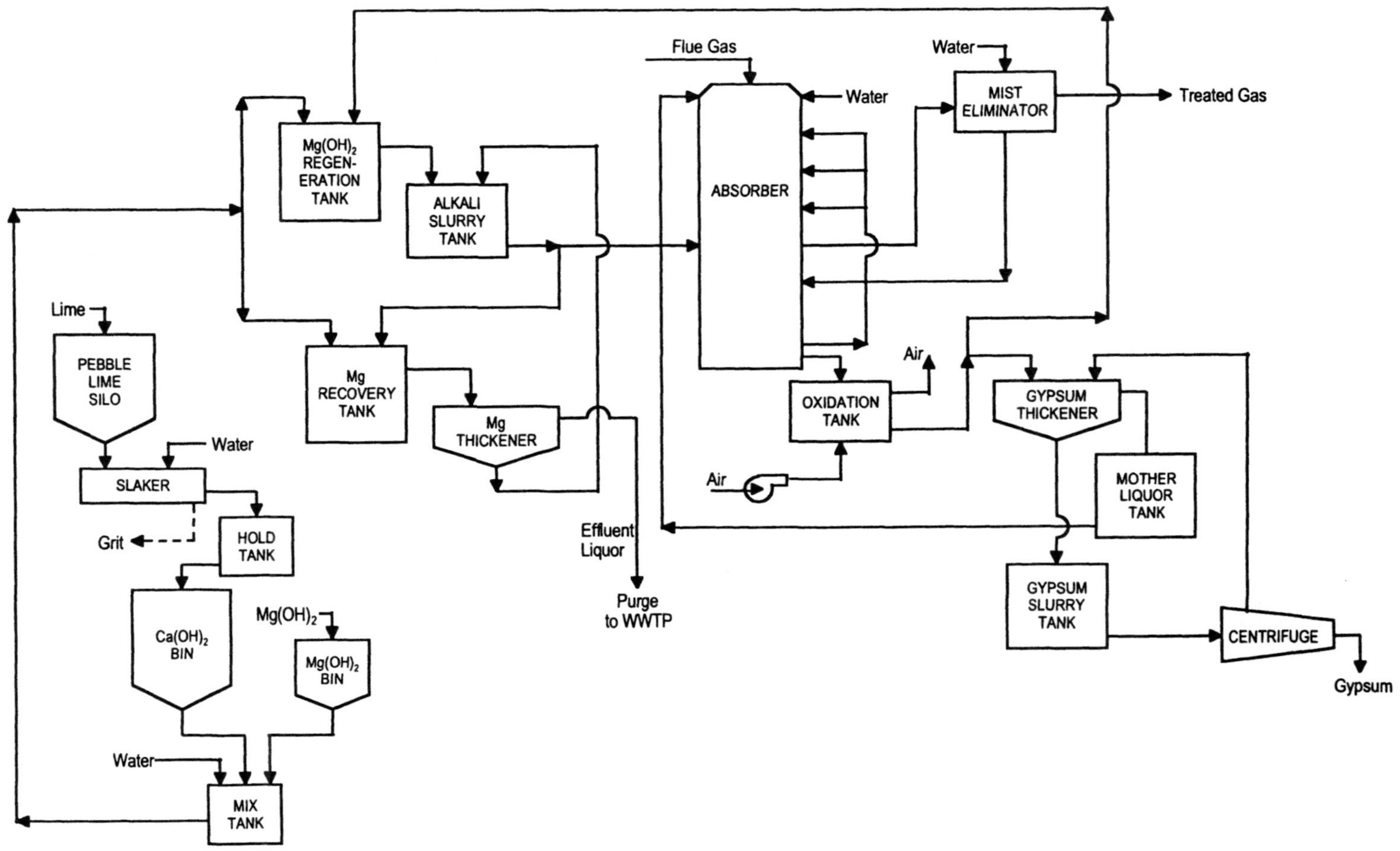

Ammonia Scrubbing – Once-Through

Commercial
SO_2 is absorbed in a solution of ammonium sulfate/sulfite after which spent solution is neutralized and, in one process, reacted with H_2S to form ammonium thiosulfate, then concentrated as a liquid fertilizer byproduct.

Process Characteristics
SO_2 Sorbent: Ammonium sulfate/sulfite solution
Principal Raw Materials: Ammonia
Potentially Saleable Byproducts: Solution of mixed ammonium sulfite/sulfate salts or thiosulfate
Solid Wastes: None
Liquid Effluents: None
Added Gaseous Emissions: Potential for "blue plume" in stack gas discharge
Inlet SO_2 Levels: up to ~4,000 ppm (Commercial Operation)
SO_2 Removal: Typical Design Range: 85-95%; Maximum Achievable at Highest SO_2: ~97%
NO_x Removal: Nil
Particulate Removal: Upstream decoupled particulate control required

Commercial Application
Number/Types: Several: Claus plant; oil-fired boilers
Locations: US; Far East
First Deployment: Late 1970s
Current Status: Active

Process Description
There are two basic process variations – one which produces a mixed ammonium sulfite/sulfate liquor byproduct and another that produces ammonium thiosulfate. The latter was specifically developed for applications where H_2S gas is available, such as in Claus plant tail gas treatment. In this approach, flue gas is contacted in a series of three absorption towers producing a spent solution of ammonium sulfite, bisulfite and a small amount of sulfate. The spent solution is then reacted with ammonia and H_2S from the Claus plant to produce an ammonium bisulfate solution. The ammonium bisulfate solution is then concentrated in an evaporator to yielding a concentrated byproduct liquor. In other once-through ammonia processes, the spent solution is simply neutralized with ammonia and, perhaps, concentrated.

Advantages/Disadvantages

Advantages
- Produces a fertilizer byproduct
- High tolerance for sorbent oxidation
- Fairly high SO_2 removal capability

Disadvantages/Limitations
- Liquid (mixed) fertilizer may have limited market value
- Potential for "blue plume" opacity issues requires very high mist removal or a wet ESP
- Need to handle ammonia, a highly regulated hazardous chemical

Principal Suppliers

Supplier	Process Byproduct
Coastal Chem	Ammonium thiosulfate
Tampella	Mixed ammonium sulfite/sulfate
Ube	Mixed ammonium sulfite/sulfate

Byproduct Processes

ATS Process — Ammonia Scrubbing — Once-Through   Commercial

Ammonia Scrubbing with Oxidation

Commercial
SO_2 is absorbed in a solution of ammonium sulfate/sulfite, the sulfite is oxidized to sulfate and ammonium sulfate is recovered as solid or solution fertilizer byproduct.

Process Characteristics
SO_2 Sorbent: Ammonia
Principal Raw Materials: Ammonia
Potentially Saleable Byproducts: Ammonium sulfate fertilizer (solution or dry)
Solid Wastes: None
Liquid Effluents: None
Added Gaseous Emissions: Potential for "blue plume" in stack gas discharge
Inlet SO_2 Levels: 1,200 to 6,000 ppm (Commercial Operations)
SO_2 Removal: Typical Design Range: 95%; Maximum Achievable at Highest SO_2: ~98%
NO_x Removal: Nil
Particulate Removal: Upstream decoupled, dry particulate control required for fertilizer byproduct

Commercial Application
Number/Types: Numerous: Mostly coal-fired boilers (only a few remain operational)
Locations: Far East; Europe; US
First Deployment: 1980s
Current Status: Active

Process Description
There are two basic versions of the process. One involves oxidation of spent solution outside the absorber; the other uses in-situ oxidation in the absorber. The latter is described, since it is most recent and has concepts likely to be used in new systems. After dry particulate removal, hot flue gas enters a prescrubber where it is contacted with a slurry saturated in ammonium sulfate. Water evaporation cools and saturates the gas and causes crystallization of ammonium sulfate. No ammonia is added so the prescrubber operates at low pH with no SO_2 removal. After demisting, the gas enters a multi-level, high L/G, countercurrent spray tower that operates as both an SO_2 absorber and oxidizer. Air and ammonia are sparged into solution collected at the bottom of the tower. Air oxidizes sulfite to sulfate and ammonia controls pH. Spent solution from the absorber is fed to the prescrubber. Slurry from the prescrubber is continuously withdrawn for dewatering, drying and product preparation. Gas from the absorber is discharged either through a high efficiency demister or a wet ESP to minimize carryover of mist that can lead to "blue plume" (opacity) problems. Blue plumes have plagued many ammonia scrubbing systems, leading to shut down of all but a few. Recent design advances including better chemistry control especially in combination with use of high efficiency demisters or a wet ESP should eliminate such problems. Pelletizing and crystallization processes are available for fertilizer-grade ammonium sulfate preparation, the later similar to that used in Caprolactam production.

Advantages/Disadvantages

<u>Advantages</u>
- Produces dry ammonium sulfate fertilizer
- No solid wastes or liquid effluents
- High SO_2 removal capabilities at high inlet SO_2

<u>Disadvantages/Limitations</u>
- Upstream particulate removal is required
- Potential for "blue plume" opacity issues requires very high mist removal or a wet ESP
- Need to handle ammonia, a highly regulated hazardous chemical

Principal Suppliers

<u>Suppliers of Commercial Processes</u>
ABB
Ishikawajima-Harima Heavy Industries (IHI)
Marsulex
Lentjes-Bischoff - owns Krupp technology

<u>Other Processes Marketed or in Development</u>
Benetech
Drager-Energie-Technik
Hoogovens (Marsulex Licensee)
Kruger I. A/S
Thyssen

Byproduct Processes

Ammonia Scrubbing with Oxidation

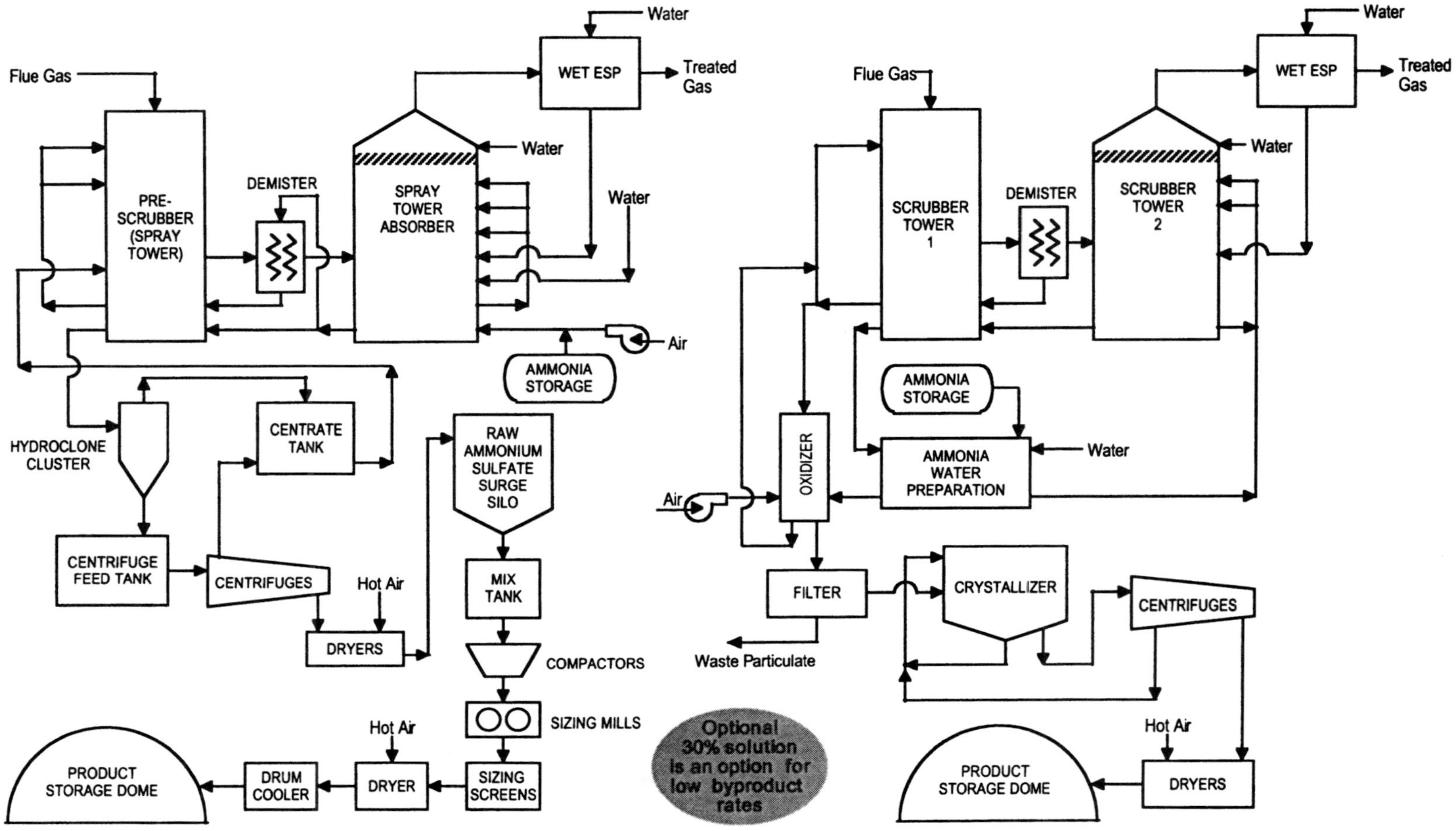

Version 1 (GEESI Process) — Commercial

Version 2 (Krupp-Koppers Process) — Commercial

Optional 30% solution is an option for low byproduct rates

Walther Process with NO$_X$ Control – Ammonia Scrubbing with SCR

Commercial
SO$_2$ is absorbed in a solution of ammonium sulfate/sulfite then NO$_X$ is reduced by SCR. Spent sulfite solution is oxidized to sulfate and ammonium sulfate is recovered as dry fertilizer product by crystallization and drying.

Process Characteristics
SO$_2$ Sorbent: Ammonia
Principal Raw Materials: Ammonia; Catalyst (for SCR)
Potentially Saleable Byproducts: Dry ammonium sulfate fertilizer
Solid Wastes: None
Liquid Effluents: None
Added Gaseous Emissions: None
Inlet SO$_2$ Levels: ~1,000 ppm (Commercial Operations)
SO$_2$ Removal: ~93% (Commercial Operations)
NO$_x$ Removal: ~85% (Commercial Operations)
Particulate Removal: Upstream decoupled, dry particulate control required for fertilizer byproduct

Commercial Application
Number/Types: One: Coal-fired boiler (Shut down in 1998 due to conversion of plant to gas)
Locations: Europe
First Deployment: 1991
Current Status: Active

Process Description
The Walther process enhanced with integrated NO$_x$ control was applied to the Karlsruhe plant in Germany. The design drew upon prior experience and design concepts used in two previous installations for SO$_2$ control only similar to processes described in Ammonia with Oxidation technology. The principal difference was that a selective catalytic reduction (SCR) unit for NO$_X$ control was added downstream of the SO$_2$ absorber. In this configuration, flue gas first enters a two-stage SO$_2$ absorption system where it is contacted with a solution of ammonium sulfite/sulfate. After demisting, the gas passes through a gas-to-gas heat exchanger and support heater after which additional ammonia is added and the gas enters an SCR unit. Gas from the SCR is then used to preheat incoming gas before being discharged. In the Karlsruhe plant, a wet ESP is used to ensure elimination of a blue plume associated with ammonium sulfate particulate. Spent scrubber solution is passed through an air blown oxidizer to convert the sulfite to sulfate after which it is sent to crystallization, dewatering, drying and product preparation. The incorporation of SCR in the gas treatment train after the ammonia SO$_2$ absorber allows control of ammonia slip from the absorber and, therefore, mitigation of potential for opacity ("blue plume") problems.

Advantages/Disadvantages

Advantages
- Produces dry ammonium sulfate fertilizer
- No solid wastes or liquid effluents
- High SO$_2$ removal capabilities at high inlet SO$_2$
- Minimization of "blue plume" potential through post scrubbing SCR

Disadvantages/Limitations
- Upstream particulate removal is required
- Need to handle ammonia, a highly regulated hazardous chemical

Principal Suppliers
Lentjes-Bischoff – Walther technology now owned by Lentjes-Bischoff through purchase of Krupp

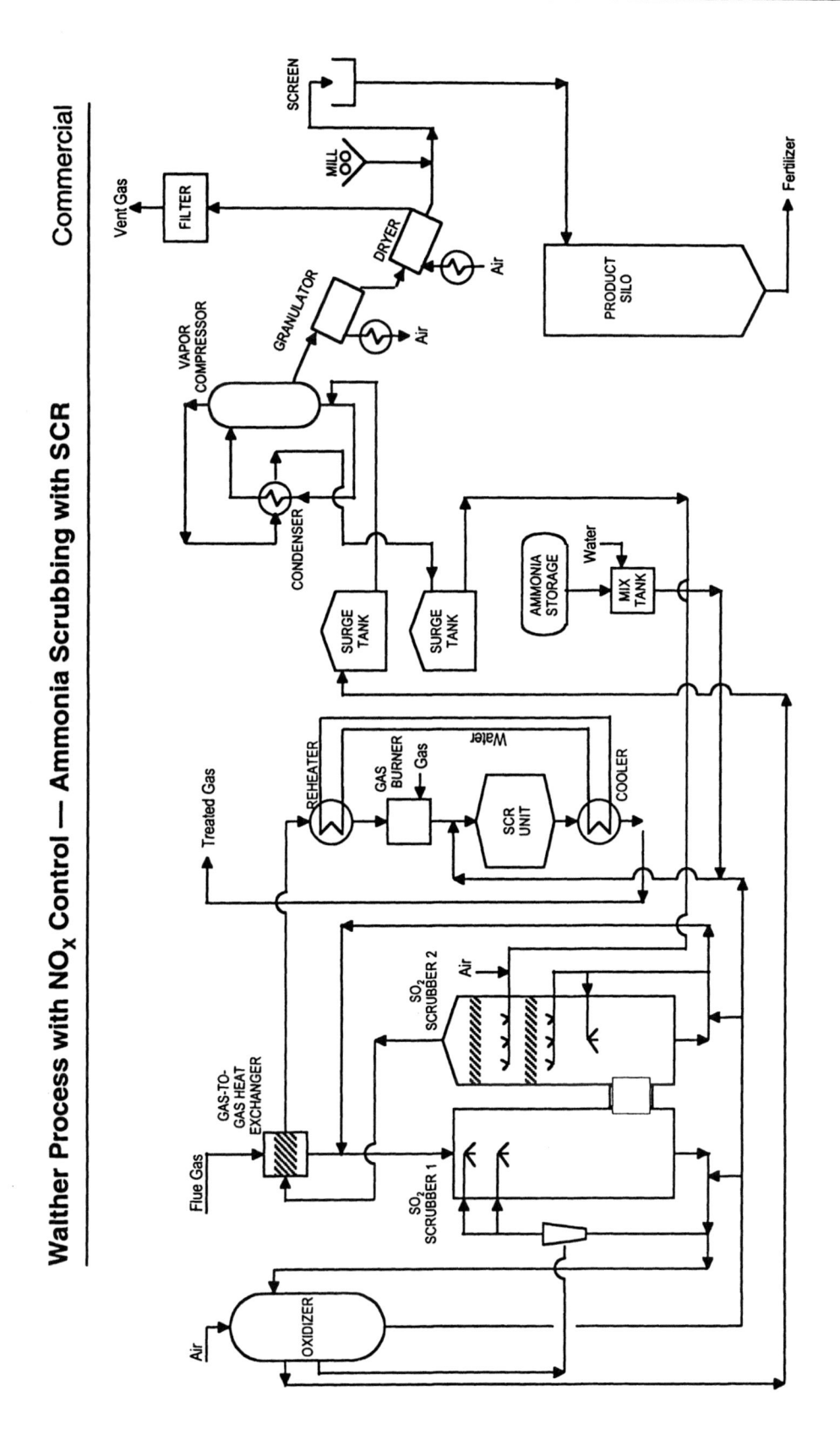

Walther Process with NO$_x$ Control — Ammonia Scrubbing with SCR   Commercial

Electron Beam Irradiation

Commercial
Flue gas is mixed with ammonia, then exposed to a high energy flux of electrons which converts SO_x and NO_x to ammonium sulfate and nitrate particulate that is collected in an ESP or fabric filter for sale as a fertilizer.

Process Characteristics
SO_2 Sorbent: Ammonia
Principal Raw Materials: Ammonia
Potentially Saleable Byproducts: Mixed ammonium sulfate and nitrate fertilizer
Solid Wastes: None
Liquid Effluents: None
Added Gaseous Emissions: None
Inlet SO_2 Levels: 1,200 to 2,500 ppm (Demonstration and Commercial Operations)
SO_2 Removal: Typical Design Range: 80-95%; Maximum Achievable at Highest SO_2: ~95%
NO_x Removal: >80% Capability (Typical 40-80%)
Particulate Removal: Upstream decoupled, dry particulate control required for fertilizer byproduct

Commercial Application
Number/Types: Two: Coal-fired boilers (one operating & one under construction)
Locations: China
First Deployment: 1997
Current Status: Active

Process Description
Several processes have been under development dating to the early 1970s, with research primarily being supported in Japan. The processes differ primarily in the manner in which the gas is exposed to the high energy flux. The Ebara E-Beam® process has been under extensive development for over twenty years and is the only one to have achieved commercial status. In the Ebara process, flue gas is first partially saturated and cooled with water to a temperature of 150°±10°F. It is then mixed with ammonia and passed to the reactor where it is subjected to beams of high energy electrons. SO_2 and NO_x are oxidized forming ammonium sulfate and nitrate particulate. The particulate are then collected downstream in an ESP or fabric filter and transferred to a storage silo. It can be sold directly as fertilizer or granulated first. The degree of SO_2 and NO_x removal can be adjusted by power application. In the first commercial application on a high sulfur coal-fired application in China, SO_2 removal is limited to 80% and NO_x removal to 40-50% by "detuning" the power to about 1.5% of the total generator capacity. Increasing the power to about 2% of the steam generator capacity for a high sulfur coal application increases SO_2 removal to about 90% and NO_x removal to 60-70% removal.

Advantages/Disadvantages

Advantages
- Produces fertilizer supplements directly
- Converts NO_x to usable byproduct rather than nitrogen
- No solid wastes or liquid effluents and no slurries to deal with
- Stack gas requires no reheat

Disadvantages/Limitations
- Upstream particulate removal must reduce heavy metal levels if byproduct to be acceptable as a fertilizer
- Relatively high energy consumption
- Need to handle ammonia, a highly regulated hazardous chemical

Principal Suppliers/Developers

Commercialized Processes
Ebara (E-Beam)

Developmental Processes
ENEL (Pulse Energization)
Karlsruhe (Electron Streaming)

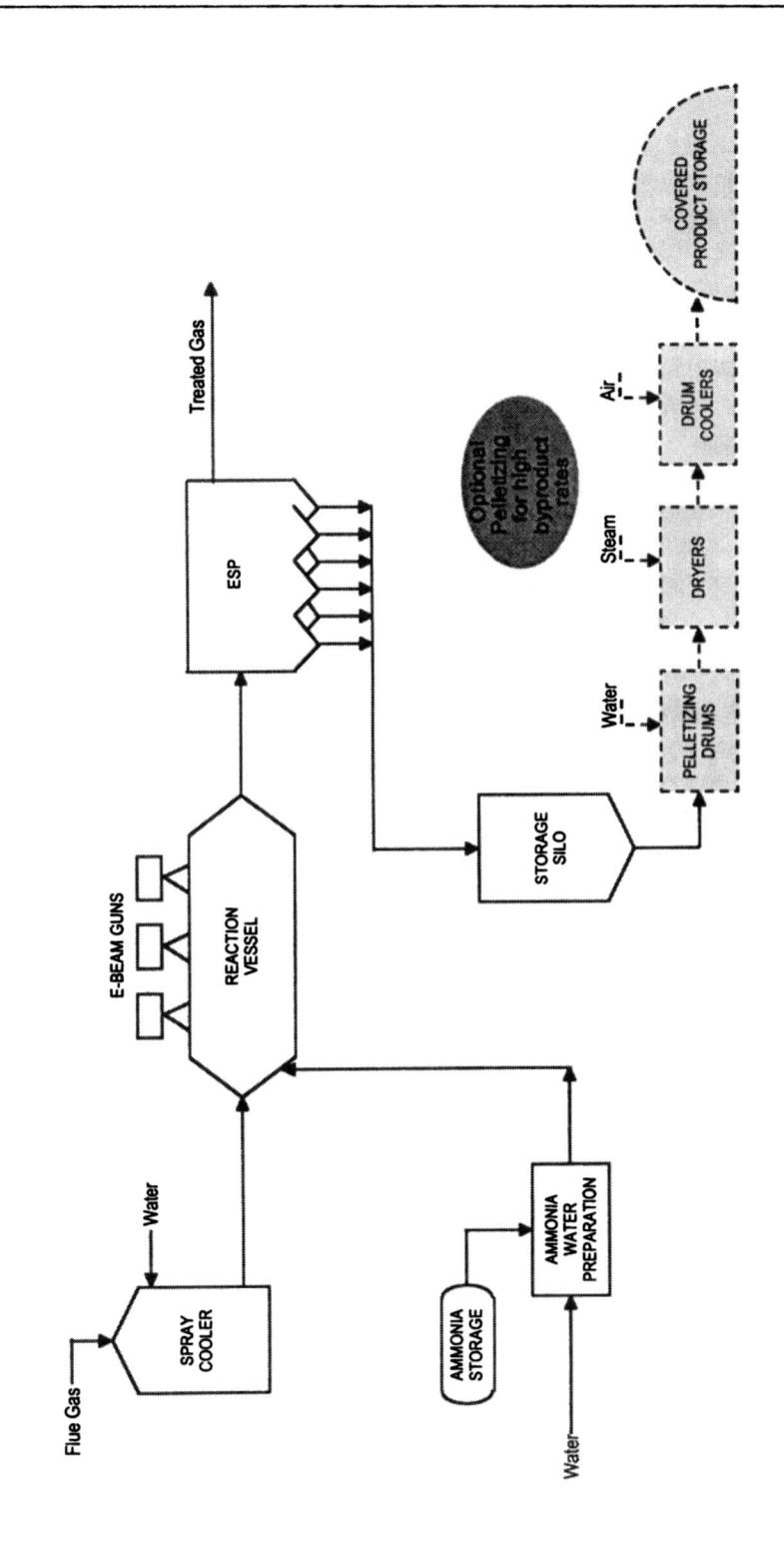

Byproduct Processes

Electron Beam Process

Ammonia Scrubbing with Acid Regeneration – Cominco Process

Commercial
SO$_2$ is absorbed in a solution of ammonium sulfate/sulfite followed by acidification with sulfuric acid, then air stripping to release SO$_2$ for conversion to acid and produce a byproduct ammonium sulfate solution.

Process Characteristics
SO$_2$ Sorbent: Ammonia
Principal Raw Materials: Ammonia; Sulfuric acid
Potentially Saleable Byproducts: SO$_2$-rich gas for conversion to sulfuric acid; Ammonium sulfate solution
Solid Wastes: None
Liquid Effluents: None
Added Gaseous Emissions: Potential for "blue plume" in stack gas discharge
Inlet SO$_2$ Levels: 7,000 to 55,000 ppm (Commercial Operations)
SO$_2$ Removal: Typical Design Range: 85-95%; Maximum Achievable at Highest SO$_2$: ~97%
NO$_x$ Removal: Nil
Particulate Removal: Upstream decoupled particulate control required

Commercial Application
Number/Types: Several: Smelters; acid plants
Locations: US; Canada
First Deployment: 1940s
Current Status: Active

Process Description
The Cominco process was developed in the 1940s specifically for application to smelter and acid plant tail gases. Thus it has been applied to very high inlet SO$_2$ levels. The gas is contacted in a multi-stage packed tower (wood slat packing was used in all the early units). Liquor is separately collected and recycled through each stage to maintain proper pH control and counter current operation for optimum SO$_2$ removal. It is also necessary to control the absorption temperature both to minimize ammonia loss and maintain favorable absorption equilibria. In smelter applications, the high SO$_2$ levels necessitate recirculation of the absorbent solution through coolers to remove the heat of reactions. This is generally not required with levels on the order of 1.0% or lower. Spent solution withdrawn from the absorber is sent to one of a set of batch tanks in which it is first acidified with 93% H$_2$SO$_4$ converting the solution to ammonium bisulfate and releasing SO$_2$ then neutralized with ammonia. The neutralized solution, now saturated in SO$_2$, is passed through an air stripping tower, producing an SO$_2$-rich air stream for feed to an acid plant and a 40% ammonium sulfate solution.

Advantages/Disadvantages

Advantages
- Produces SO$_2$-rich gas for conversion to sulfuric acid and byproduct ammonium sulfate solution
- No solid wastes
- Relatively high SO$_2$ removal capabilities at high inlet SO$_2$

Disadvantages/Limitations
- Upstream particulate removal is required
- Potential for "blue plume" opacity issues requires very high mist removal or a wet ESP
- Need to handle ammonia, a highly regulated hazardous chemical

Principal Suppliers
Cominco

Byproduct Processes

Ammonia Scrubbing with Acid Regeneration — Cominco Process   Commercial

Magnesium Oxide Recovery Process

Commercial
SO_2 is absorbed in a slurry of magnesium hydroxide forming magnesium sulfite/bisulfite which is then thermally regenerated under reducing conditions to produce an SO_2 offgas most suitable for conversion to sulfuric acid.

Process Characteristics
SO₂ Sorbent: Magnesium hydroxide/sulfite slurry
Principal Raw Materials: Magnesium hydroxide; Fuel gas/oil for thermal regenerator
Potentially Saleable Byproducts: SO_2-rich gas most suitable for conversion to sulfuric acid
Solid Wastes: None
Liquid Effluents: None
Added Gaseous Emissions: Thermal regenerator combustion offgas
Inlet SO_2 Levels: 1,000 to 2,500 ppm (Commercial Operation)
SO_2 Removal: 90-95% (Commercial Operation)
NO_x Removal: Nil
Particulate Removal: Upstream decoupled, dry particulate control generally required

Commercial Application
Number/Types: Several: Utility coal and oil-fired boilers
Locations: US; Europe; Japan
First Deployment: Mid-1970s (Current design concepts)
Current Status: Active (But not proactively marketed)

Process Description
Flue gas is first prescrubbed to remove impurities that can contaminate the absorption circuit (e.g., chlorides) and residual particulate. Cooled and humidified gas is then contacted with a slurry of magnesium hydroxide and sulfite solids in a buffered solution of magnesium sulfite/bisulfite/sulfate which absorbs SO_2. (Note – magnesium salts are much more soluble than equivalent calcium salts.) The scrubbing slurry is continuously replenished with a mix of fresh and regenerated magnesium hydroxide. Spent slurry is continuously withdrawn. The bleed slurry is dewatered and then dried. At this point the system can be completely decoupled between the scrubbing and regeneration systems. Solids are generally stored in a spent solids silo to allow semi-independent control of the regeneration circuit. From the spent sorbent silo, the solids are transported to the regeneration system where the solids are calcined with coal, coke, fuel oil or natural gas. A fluid bed is usually used for thermal regeneration. During regeneration SO_2 is released regenerating magnesium oxide. Under reducing conditions, magnesium sulfate is also regenerated releasing SO_2. The SO_2-rich gas is usually on the order of 7-10% — suitable for conversion to sulfuric acid, but too dilute for economic recovery to sulfur. Regenerated sorbent from the calciner is returned to the scrubbing system. A slaking system is required to hydrate the regenerated sorbent prior to introduction to the scrubber.

Advantages/Disadvantages

Advantages
- Produces SO_2-rich byproduct gas for conversion to sulfuric acid
- No significant solid wastes or liquid effluents
- Scrubbing and regeneration/acid production can be effectively decoupled
- High tolerance for sorbent oxidation (wide range of SO_2 levels)

Disadvantages/Limitations
- Relatively complicated process configuration involving movement of solids
- Potential for high sorbent attrition losses
- Process geared more for acid production than elemental sulfur
- Lack of recent applications may limit technology "know how"

Principal Developers/Suppliers
Drager-Energie-Technik
Lentjes-Bischoff (Bischoff)
Marsulex

Mitsui
Onahama-Tsukishima
Raytheon Engineers

(United Engineers & Constructors)/PECO

Byproduct Processes

Magnesium Oxide Recovery Process

Commercial

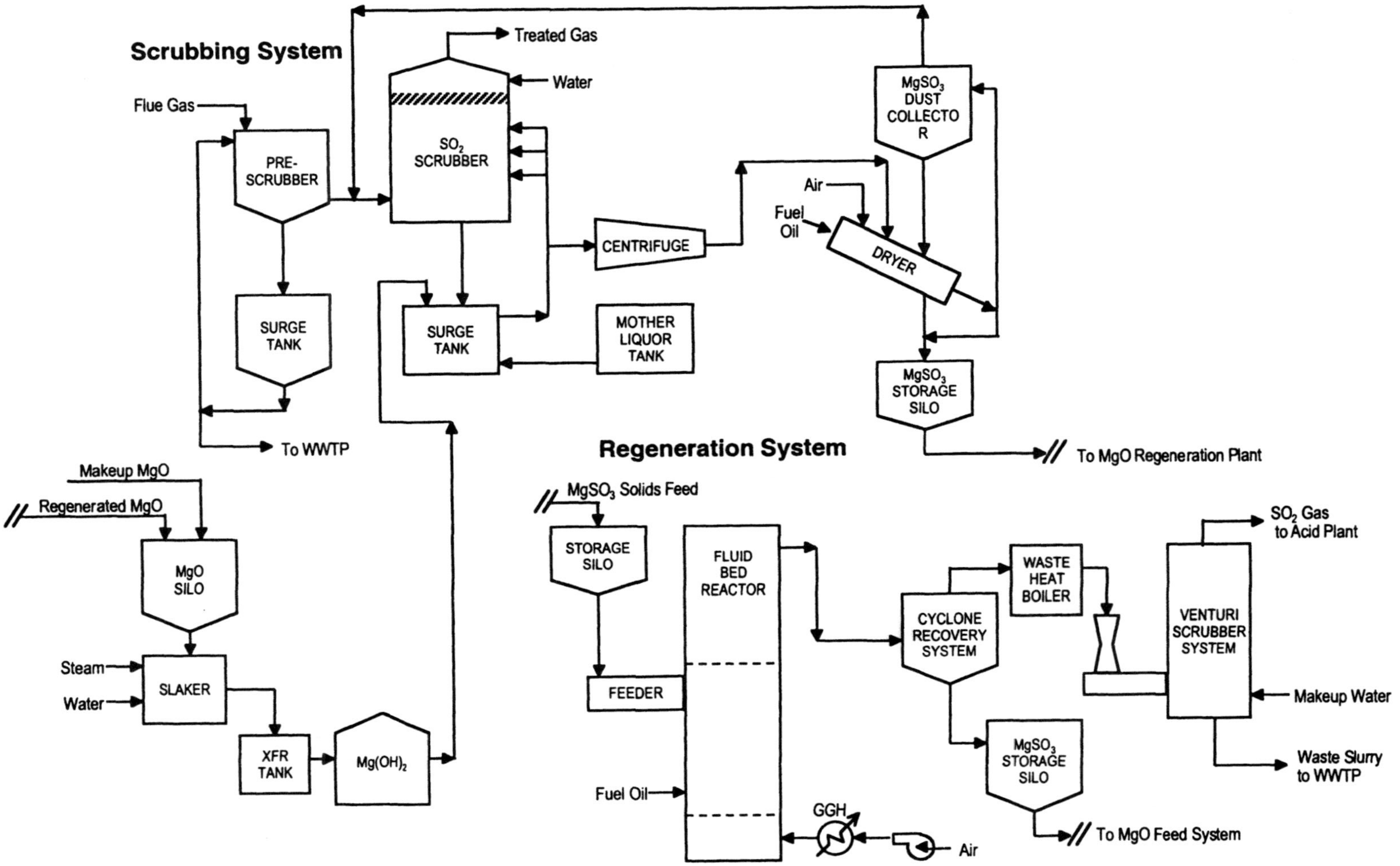

Scrubbing System

Regeneration System

Direct Sulfuric Acid Conversion

Commercial
SO$_2$ in hot, particulate-free flue gas is oxidized to SO$_3$ in a gas phase catalytic converter. The gas is cooled and SO$_3$ is absorbed in sulfuric acid. Two processes have been commercialized – Cat-Ox & WSA (SNOX precursor).

Process Characteristics
SO$_2$ Sorbent: Sulfuric acid
Principal Raw Materials: Catalyst; Natural gas (support fuel for trim heaters, as required)
Potentially Saleable Byproducts: Sulfuric acid (93-98%)
Solid Wastes: Spent catalyst
Liquid Effluents: None
Added Gaseous Emissions: None
Inlet SO$_2$ Levels: 1,000 to 60,000 ppm (Commercial Operations)
SO$_2$ Removal: Typical Design Range: 90-95%; Maximum Achievable at Highest SO$_2$: ~95%
NO$_x$ Removal: Nil
Particulate Removal: Upstream decoupled, highly efficient, hot dry particulate control is required

Commercial Application
Number/Types: Many: Combustion boilers; refineries; acid plants; roasters; chemical plants
Locations: North America (Cat-Ox); Europe and Asia (WSA)
First Deployment: 1970s (Cat-Ox); 1980 (WSA)
Current Status: Active

Process Description
These processes require hot gas to effect the conversion reaction; and low particulate loadings to minimize buildup of particulate in the SO$_2$ converter catalyst bed, which retains most all of the particulate carry-through. Therefore, in a typical boiler application, either a hot side, high efficiency particulate control device is required (e.g., hot-side electrostatic precipitator) or regenerative gas-to-gas heat exchangers. In either case, natural gas-fired support heaters would normally be required to provide additional heat. In a typical boiler application, a hot-side ESP would be used. Then SO$_2$ in the flue gas would be converted to SO$_3$ in a catalytic converter at 900°-950°F. Gas exiting the converter would be cooled to about 450°F first in an economizer (to heat boiler feed water) and then a regenerative exchanger (to heat combustion air). The temperature limit on the cooling is to remain above the dew point of acid which limits corrosion to allow expensive materials of construction. (Corrosion was a major problem in early testing and the first boiler application.) The gas then enters a contactor where the SO$_3$ is absorbed in a cool stream of recirculated sulfuric acid. Sulfuric acid is continuously withdrawn from the recirculating stream and further cooled prior to storage.

Advantages/Disadvantages

Advantages
- Produces sulfuric acid directly
- Net steam production above ~3,000 ppm SO$_2$
- Low energy consumption esp. at high inlet SO$_2$
- Low particulate emissions (could be offset by costs for high efficiency upstream collection and SO$_2$ converter catalyst bed replacement)

Disadvantages/Limitations
- Relatively low sulfuric acid concentration (90-95%) and potential for some contamination (e.g., chlorides) may limit acid uses
- Possible high SO$_2$ converter catalyst replacement due to particulate pluggage
- High levels of corrosion plagued early applications of the Cat-Ox system and dampened enthusiasm for this process

Principal Suppliers
Haldor Topsoe (WSA and SNOX Processes)
Monsanto EnviroChem (Cat-Ox Process)

Byproduct Processes

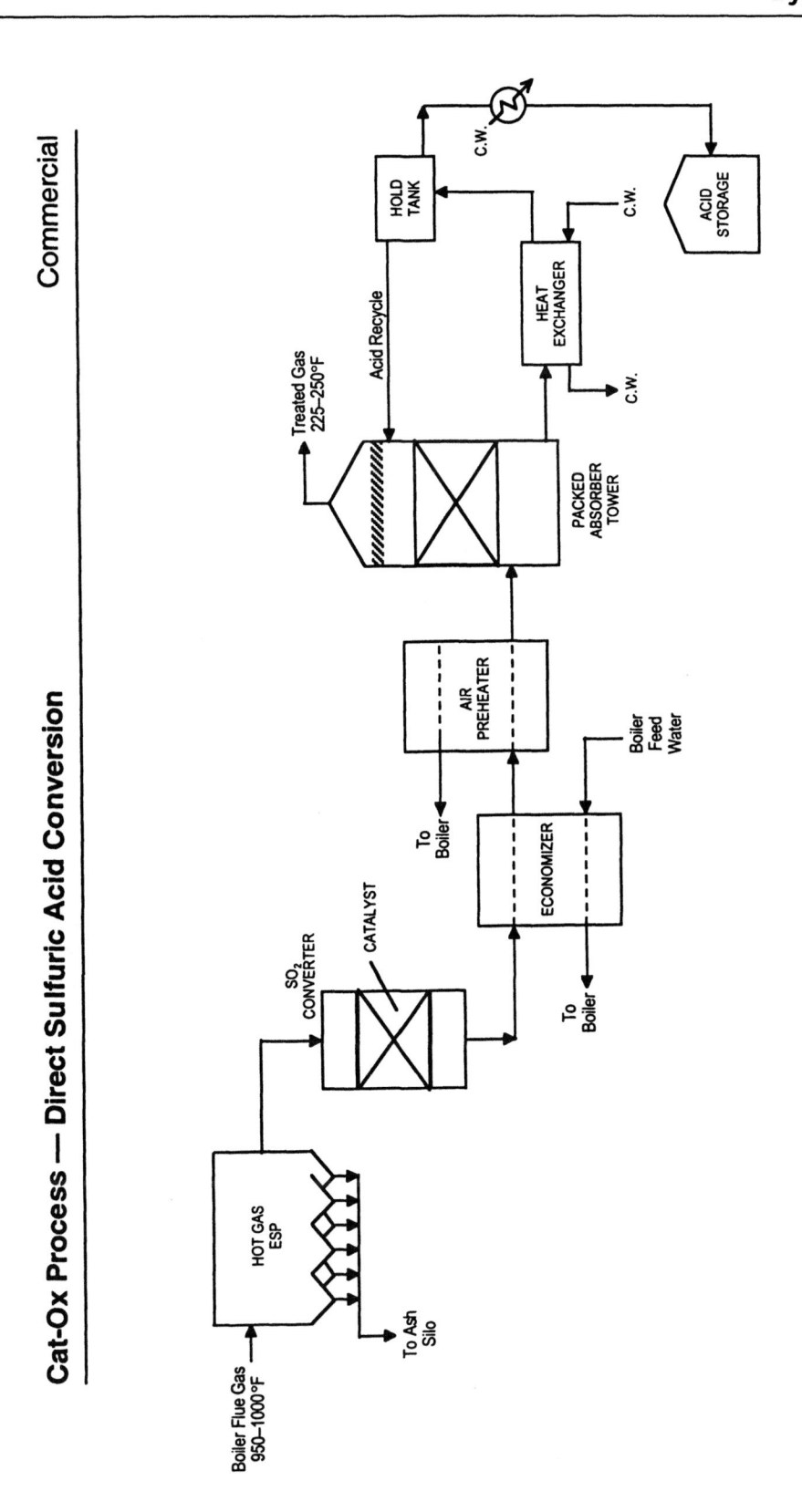

Cat-Ox Process — Direct Sulfuric Acid Conversion

Commercial

Direct Sulfuric Acid Conversion with Integrated NO$_x$ Control

Commercial
The same as the direct sulfuric acid (e.g., WSA) process, but with integrated NO$_x$ control by selective catalytic reduction.

Process Characteristics
SO$_2$ Sorbent: Sulfuric acid
Principal Raw Materials: Ammonia (for NO$_x$ reduction); Catalysts (SCR and acid conversion); Natural gas (support fuel for trim heaters, as required)
Potentially Saleable Byproducts: Sulfuric acid (93-95%)
Solid Wastes: Spent catalysts
Liquid Effluents: None
Added Gaseous Emissions: None
Inlet SO$_2$ Levels: 600 to 55,000 ppm (Commercial Operations)
SO$_2$ Removal: Typical Design Range: ~95%; Maximum Achievable at Highest SO$_2$: ~98%
NO$_x$ Removal: 90-95%
Particulate Removal: Upstream decoupled, highly efficient dry particulate control is required

Commercial Application
Number/Types: Several: Utility and industrial boilers; chemical plants
Locations: Europe; US; Asia
First Deployment: 1987 (SNOX); 1988 (DESONOX)
Current Status: Active

Process Description
Two similar processes involving direct production of sulfuric acid have been commercialized for integrated SO$_2$ and NO$_x$ control. They require hot gas to effect the conversion reactions and low particulate loadings to minimize pluggage of the SO$_2$ converter catalyst bed which retains most particulate carry-through. In a typical boiler application, flue gas first passes through a high efficiency particulate control device (e.g., fabric filter). It is then heated in a regenerative heat exchanger supplemented, as required, by a natural gas-fired heater to raise the gas temperature to ~700°-800°F. A mixture of air and ammonia is added and the gas passes through a selective catalytic reduction (SCR) reactor where 90-95% of the NO$_x$ is removed. After leaving the SCR, the gas is heated to 770°F and passed through the SO$_2$ to SO$_3$ catalytic converter. Gas exiting the converter is used in the regenerative exchanger to heat the incoming flue gas ahead of the SCR. This cools the SO$_3$-laden gas to about 300°F before it enters a falling film sulfuric acid condenser where it is further cooled with ambient air to about 200°F. The acid condenses on corrosive resistant (e.g., borosilicate) tubes, and is collected, cooled, conditioned and stored. Heated air from the condenser at ~400°F is used for combustion air.

Advantages/Disadvantages

Advantages
- Produces sulfuric acid directly
- Simultaneous NO$_x$ and SO$_x$ control
- Net steam production above ~2,500 ppm SO$_2$
- Low energy consumption esp. at high inlet SO$_2$
- Higher NO$_x$ removal than SCR alone with ability for removal of NH$_3$ slip in the SO$_2$ converter

Disadvantages/Limitations
- Moderately low sulfuric acid concentration (93-95%) and potential for some contamination (e.g., chlorides) may limit acid uses
- Possible high SO$_2$ converter catalyst replacement due to pluggage from particulate
- Need to handle ammonia, a highly regulated hazardous chemical

Principal Suppliers
Haldor Topsoe (SNOX Process)
Degussa/Lentjes/Lurgi (DESONOX)

Byproduct Processes

SNOX Process — Direct Sulfuric Acid Conversion with NO$_x$ Control — Commercial

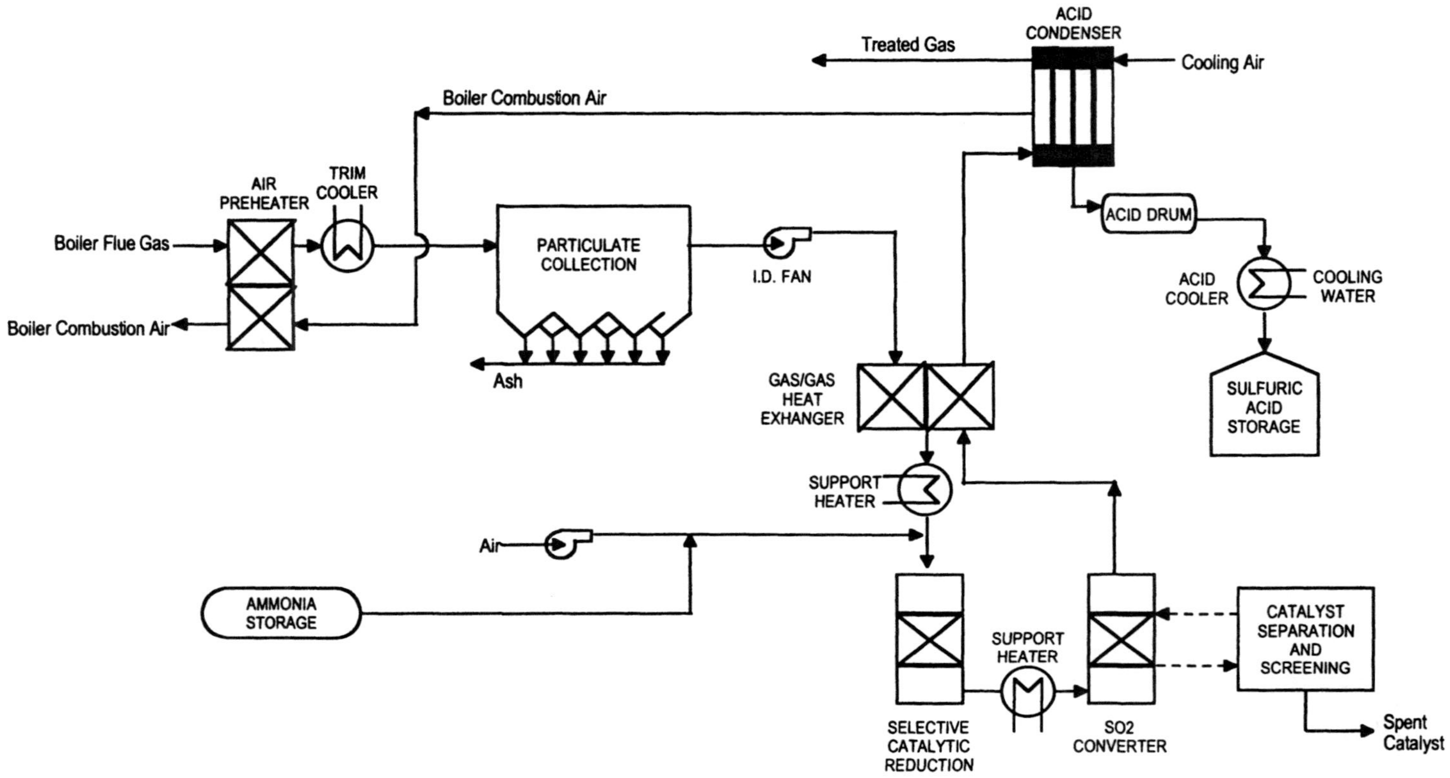

Cold Water Scrubbing with Thermal Stripping

Commercial
Cold water is used for SO$_2$ absorption followed by steam stripping of the SO$_2$ to produce an SO$_2$-rich gas for conversion to acid or condensing to liquid SO$_2$ for sale.

Process Characteristics
SO$_2$ Sorbent: Water
Principal Raw Materials: Cold water (usually seawater); Caustic for discharge water neutralization
Potentially Saleable Byproducts: SO$_2$-rich gas for conversion to acid or condensed to liquid SO$_2$
Solid Wastes: None
Liquid Effluents: Neutralized warm water
Added Gaseous Emissions: None
Inlet SO$_2$ Levels: up to 55,000 ppm
SO$_2$ Removal: Typical Design Range: ~99%
NO$_x$ Removal: None
Particulate Removal: Upstream decoupled, highly efficient dry particulate control is required

Commercial Application
Number/Types: Several: Copper and lead smelters
Locations: Europe
First Deployment: 1973
Current Status: Active

Process Description
The gas enters the bottom of an absorption tower where it is contacted with a countercurrent flow of cold water that absorbs the SO$_2$. High SO$_2$ removal efficiencies can be attained, on the order of 98-99%, however, large amounts of water are required. The amount of water is dependent upon the water temperature. For example, to achieve the same degree of removal efficiency, twice as much water is required with 20°C water as with 0°C water. In the commercial applications to date, seawater is used because of its ready availability and low temperature. Where cold seawater is not available, the process may not be cost effective. Acidic water from the bottom of the tower containing 1.0 ± 0.2% SO$_2$ is sent to a stripping tower. A surge tank is provided between the absorber and stripping column to accommodate shutdowns in the acid plant and smooth swings in absorber operation. Liquor fed to the stripping column is first preheated in a heat exchanger with degassed hot water from the stripper, then trim heated with steam to 60°C. The top of the stripping column is designed as a cooling tower to condense water vapor. For this purpose, about 15% of the absorber tower blowdown is fed directly to the top of the column, bypassing the heat exchanger system. SO$_2$-rich gas from the stripping column can either be sent to the acid plant or converted to liquid SO$_2$. For liquid SO$_2$, the gas is passed to a two-step sulfuric acid drying tower and then to a precooler after which the SO$_2$ is condensed.

Advantages/Disadvantages

Advantages
- Produces SO$_2$-rich gas for conversion to acid or liquid SO$_2$
- No solid wastes
- No slurries or scaling/plugging problems
- High SO$_2$ removal capability

Disadvantages/Limitations
- Requires a large quantity of cold water which usually means northern sites with ready access to seawater
- Discharge of spent water may encounter environmental constraints
- Economics generally dictate a captive or local use of acid
- Probably not cost effective for low concentration gases

Principal Suppliers
Boliden

Cold Water Scrubbing with Thermal Stripping

Commercial

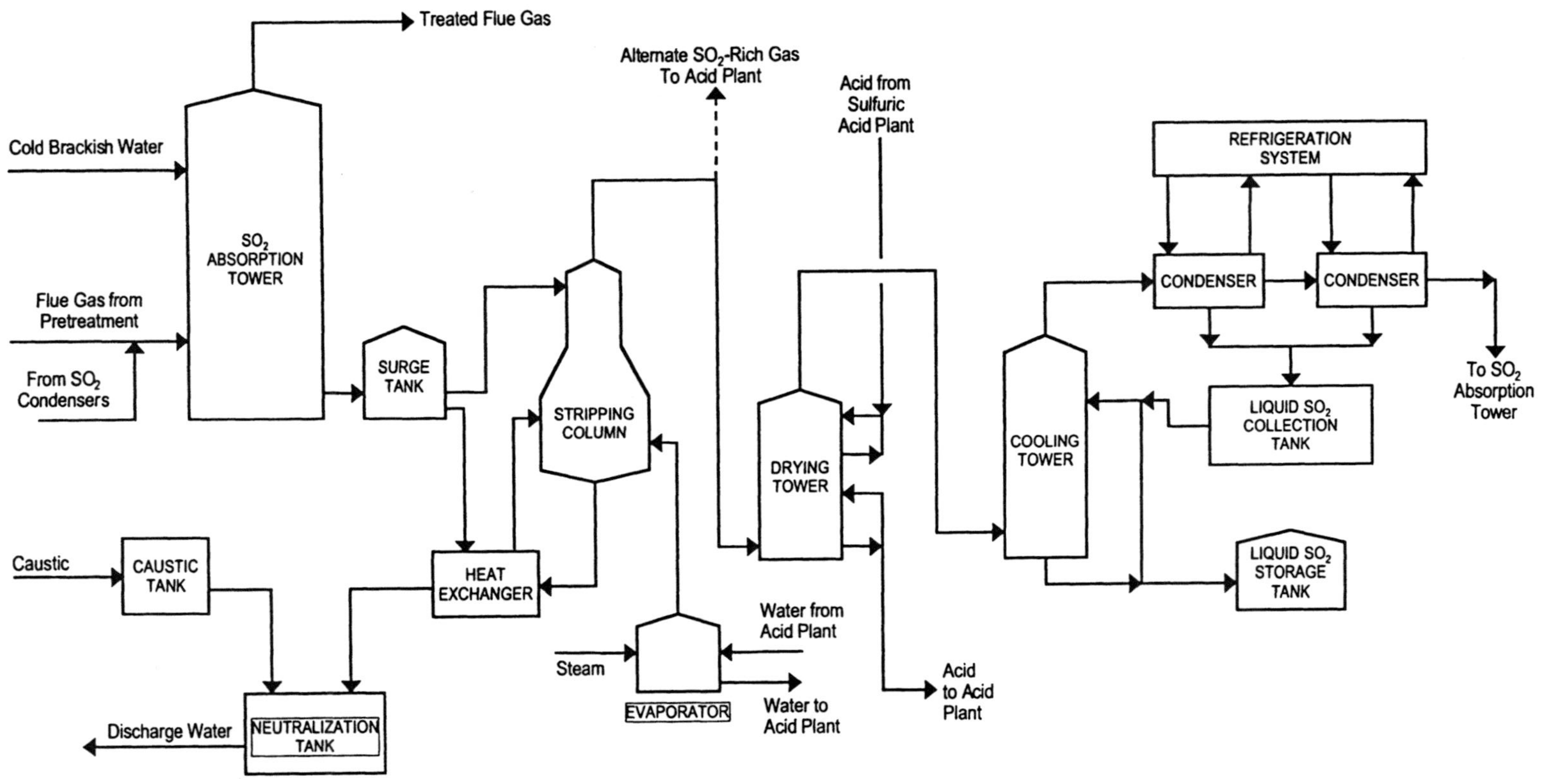

Wellman Lord Process

Commercial
Absorption of SO$_2$ in a solution of sodium sulfite/bisulfite from which absorbed SO$_2$ is thermally stripped regenerating the absorbent and producing an SO$_2$-rich gas for further processing.

Process Characteristics
SO$_2$ Sorbent: Sodium sulfite/bisulfite solution
Principal Raw Materials: Caustic or soda ash; Lime (for treating prescrubber blowdown)
Potentially Saleable Byproducts: SO$_2$-rich gas for conversion to sulfuric acid or sulfur; Sodium sulfate
Solid Wastes: Wastes from the prescrubber blowdown treatment
Liquid Effluents: Purge liquors required to control soluble impurities buildup
Added Gaseous Emissions: None
Inlet SO$_2$ Levels: 1,200 to 6,000 ppm (Commercial Operations)
SO$_2$ Removal: Typical Design Range: 90-95%; Maximum Achievable at Highest SO$_2$: 99+%
NO$_x$ Removal: Nil
Particulate Removal: Upstream decoupled particulate control is required

Commercial Application
Number/Types: Many: Utility and industrial boilers; refineries; acid plants
Locations: US; Europe; Asia
First Deployment: 1970s
Current Status: Not actively marketed

Process Description
Flue gas is first passed through a prescrubber to remove particulate and soluble impurities which are discharged through a prescrubber blowdown treatment system. The humidified gas is then contacted in a high efficiency absorber (usually a tray tower) with a solution of sodium sulfite and bisulfite which absorbs SO$_2$. Most of the spent absorbent liquor is fed to a series of multi-effect evaporator/crystallizers where SO$_2$ is evolved and sodium sulfite is crystallized. The crytallized sulfite is dissolved in condensate and returned to the absorber. A bleed stream of spent absorbent is sent to a purge crystallizer to remove sodium sulfate formed from oxidation of absorbed SO$_2$. A small purge stream is also bled from the system to control build up of impurities such as chlorides.

Advantages/Disadvantages

Advantages
- Clear solution scrubbing should minimize scaling in the absorber
- Applicable to a wide range of inlet SO$_2$
- Produces SO$_2$-rich gas for further processing
- Well-developed technology with a strong technical foundation

Disadvantages/Limitations
- Highly corrosive conditions in the regeneration and recovery circuits require expensive materials
- Scaling/plugging problems in evaporators intensify maintenance requirements
- Requires a purge stream to control impurities
- Relatively complex process configuration
- Relatively high steam requirements for absorbent regeneration
- Lack of recent applications may limit technology "know how"

Principal Suppliers
Kvaerner Process Technologies
Lurgi (Kvaerner Licensee)

Byproduct Processes

Wellman Lord Process — Commercial

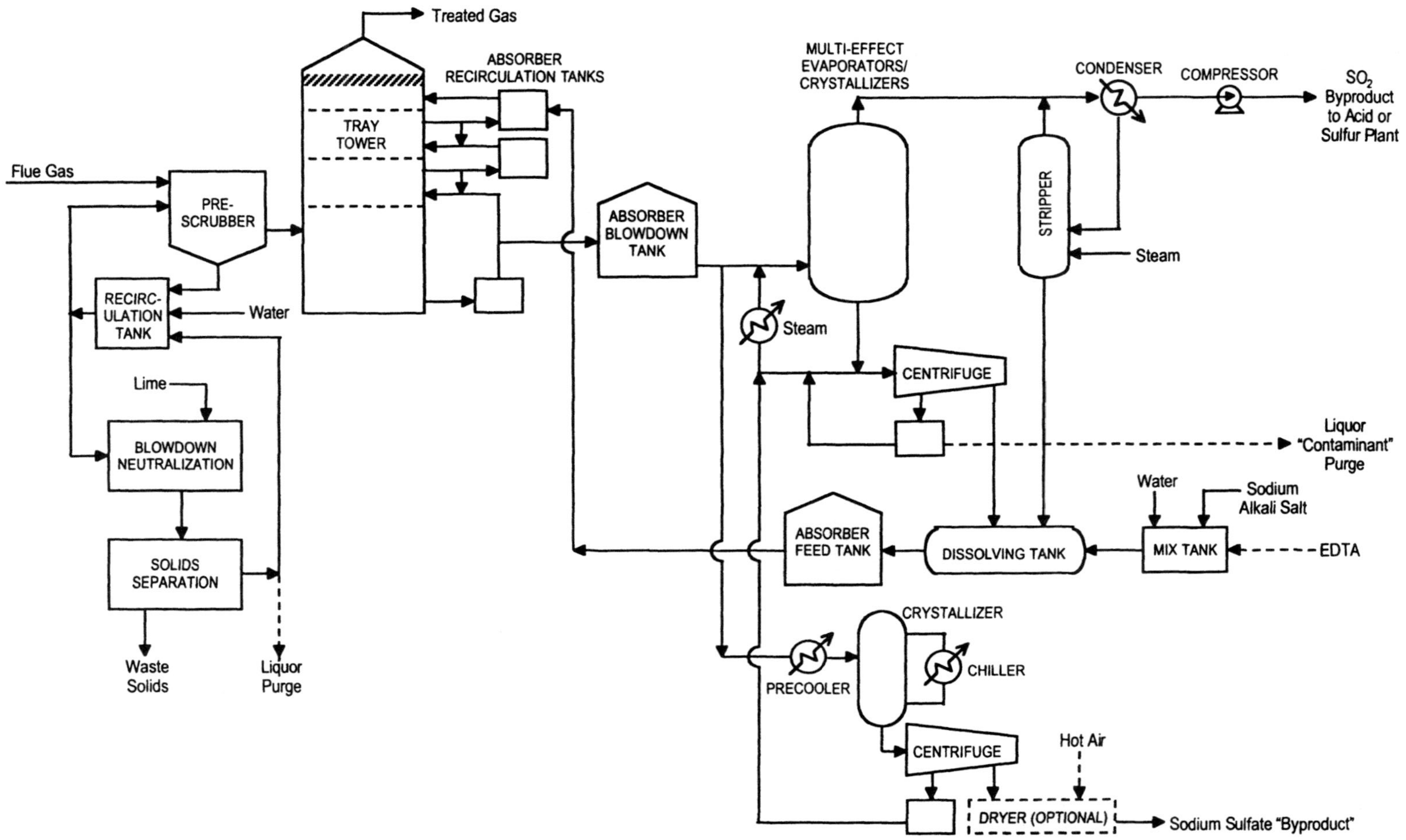

Solinox Process

Commercial
A technology based upon a conventional absorption/desorption cycle using an organic ether for SO_2 sorption, absorbent regeneration in an indirect heated stripper, and conversion of evolved SO_2-rich gas to sulfur or acid.

Process Characteristics
SO_2 Sorbent: Tetraethyleneglycol dimethylether
Principal Raw Materials: Tetraethyleneglycol dimethylether makeup; Lime for prescrubber blowdown treatment
Potentially Saleable Byproducts: SO_2-rich gas for conversion to sulfuric acid or sulfur
Solid Wastes: None
Liquid Effluents: Pretreated prescrubber blowdown liquor (with purge liquor to control impurities buildup in the solvent)
Added Gaseous Emissions: Potential for some small amount of solvent release from the water was solvent recovery absorber
Inlet SO_2 Levels: 5,000 to 20,000 ppm (Commercial Operations)
SO_2 Removal: Typical Design Range: 90-95%; Maximum Achievable at Highest SO_2: 99+%
NO_x Removal: Nil
Particulate Removal: Upstream decoupled particulate control is required

Commercial Application
Number/Types: Numerous: Fossil fuel-fired boiler; ore smelters; pulp mill; metallurgical plant
Locations: Europe
First Deployment: Late 1970s
Current Status: Active

Process Description
Flue gas is first cooled in a gas-to-gas heat exchanger with treated gas, then water scrubbed in a prescrubber to both remove residual particulate and soluble impurities that might contaminate the sorbent and further cool the gas to minimize solvent evaporation. Precooled, humidified gas is then contacted in a countercurrent, multistage absorber with tetraethyleneglycol dimethylether which physically absorbs the SO_2. Treated gas is then subjected to several stages of water wash to recover any vaporized sorbent prior to discharge through the inlet flue gas precooler exchanger. Spent solvent is passed to a conventional stripper with a steam reboiler. SO_2-rich gas from the stripper is chilled to condense water and solvent to concentrate the gas prior to conversion to elemental sulfur or sulfuric acid and recover vaporized solvent. Regenerated solvent is returned to the SO_2 absorber. The process is also capable of removal of hydrocarbons such as benzene which are quite soluble in the solvent. These can be removed from the SO_2 byproduct in an additional fractionation step or carried out with the SO_2 byproduct and destroyed by oxidation during further processing.

Advantages/Disadvantages

Advantages
- Produces SO_2 byproduct liquid or gas
- Capability for removal of hydrocarbons in the flue gas
- No scale potential in the circuits handling solvent
- Physical absorption has a high tolerance to fluctuations in inlet SO_2 compared to chemical absorption processes

Disadvantages/Limitations
- Relatively high capital investment
- Purge liquors from the prescrubber and control of impurities buildup
- Potential for solvent offgassing

Principal Suppliers
Linde A.G.

Byproduct Processes 91

SOLINOX Process — Organic Solvent with Thermal Regeneration Commercial

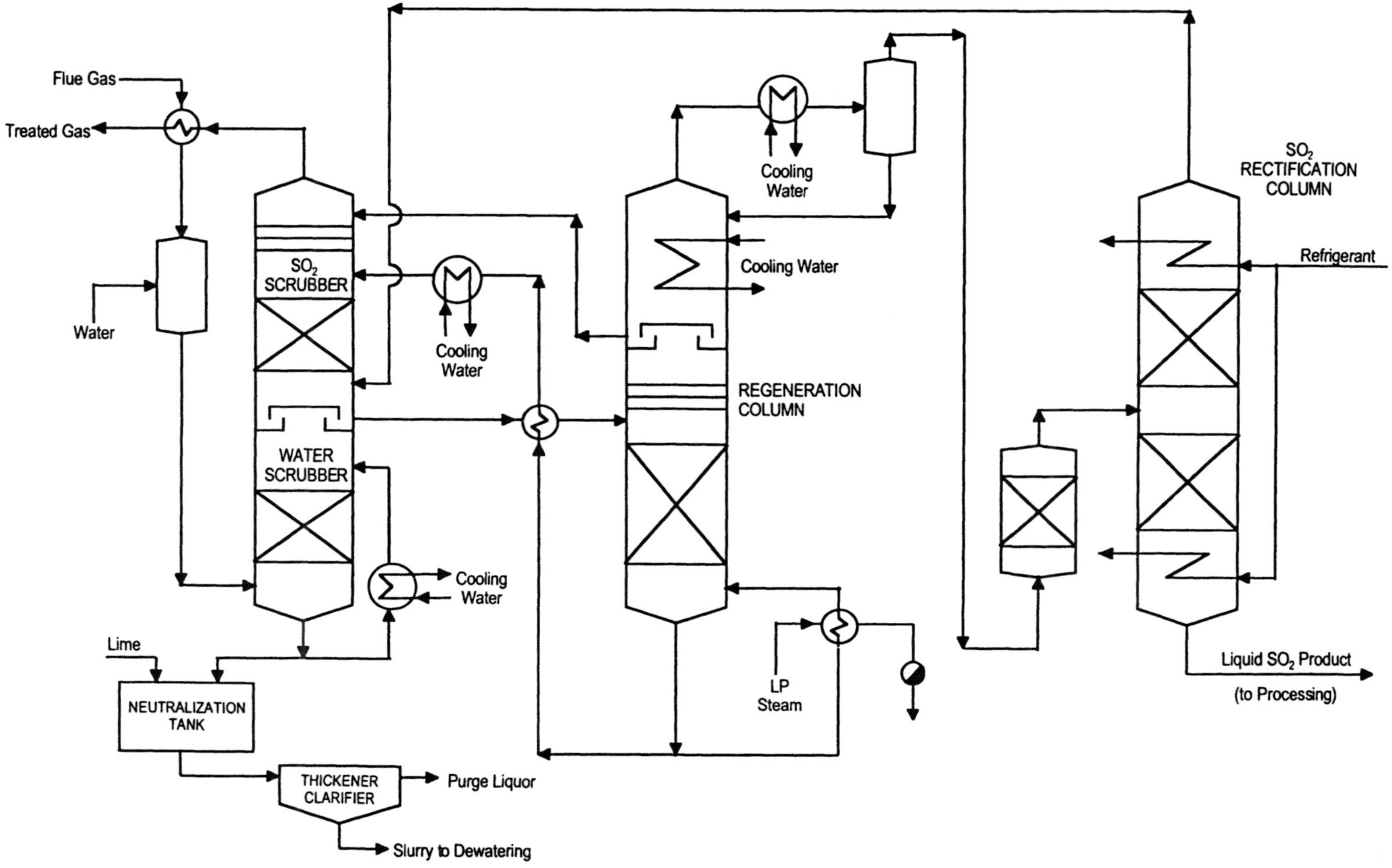

Amine Solution with Thermal Regeneration

Commercial/Developmental
Absorption of SO$_2$ in amine solution, then stripping of the SO$_2$ from the solution usually with steam, sometimes assisted by acidification, followed by conversion to acid or sulfur.

Process Characteristics
SO$_2$ Sorbent: ASARCO (Dimethylaniline) Dow & Union Carbide (Proprietary) Monsanto (Ethanolamine glutarate) Lurgi (Xylidine and toluidine in water)
Principal Raw Materials: Lime or caustic soda (prescrubber blowdown treatment); Amine makeup; Sulfuric acid (ASARCO)
Potentially Saleable Byproducts: SO$_2$-rich gas for conversion to acid or sulfur
Solid Wastes: Prescrubber blowdown wastes
Liquid Effluents: Purge liquor for soluble impurities buildup
Added Gaseous Emissions: Potential for a small amount of amine loss to the stack (in current processes, amines used are classified as nonhazardous)
Inlet SO$_2$ Levels: 3 to 8% – ASARCO and Lurgi (Commercial Operations); 1,000 to 50,000 ppm – Dow and Union Carbide (Tested)
SO$_2$ Removal: Typical Design Range: 90%; Maximum Achievable at Highest SO$_2$: 98+%
NO$_x$ Removal: Nil (But NO$_x$ absorbents are being developed for the more recent processes)
Particulate Removal: Upstream decoupled particulate control required

Commercial Application
Number/Types: Many: Metallurgical plants (inactive)
Locations: ASARCO – About ten worldwide (all shut down); Lurgi – Several, mainly in Europe (all shut down)
First Deployment: 1940s
Current Status: Inactive – All inactive but may be available for license or joint development

Process Description
Amine scrubbing processes have four basic steps. Flue gas is first contacted in a water prescrubber to remove impurities (dust, chlorides, SO$_3$, etc.) that can contaminate the absorbent solution. The humidified gas passes to a multi-stage absorber tower (usually a tray column) with interstage collectors and recycle. Spent solution is sent to a regeneration circuit where SO$_2$ is thermally stripped from the solution usually using indirect steam. In the DMA process sulfuric acid is sometimes used to assist stripping and dry the SO$_2$. A portion of the regenerated solution is sent to a salt recovery system where heat stable amines formed from oxidation of absorbed SO$_2$ as well as SO$_3$ absorption are removed.

Advantages/Disadvantages

Advantages
- Clear solution minimizes scaling and plugging problems
- Produces SO$_2$ byproduct
- Capable of handling very high inlet SO$_2$

Disadvantages/Limitations
- Low SO$_2$ applications may be uneconomic due to low tolerance to oxidation
- Relatively complex and energy intensive
- High capital costs
- Lack of recent commercial applications may limit technology "know how"

Principal Suppliers/Developers

Supplier/Developer	Process Name	Status
ASARCO	DMA	Commercial (Inactive)
Dow	Dow Process	Pilot (Inactive)
Lurgi/Gesellschaft/Metallgesellschaft	Sulphidine Process	Commercial (Inactive)
Monsanto	NOSOX	Pilot (Inactive)
Union Carbide (Turbosonic)	CANSOLV	Demonstration (Inactive)

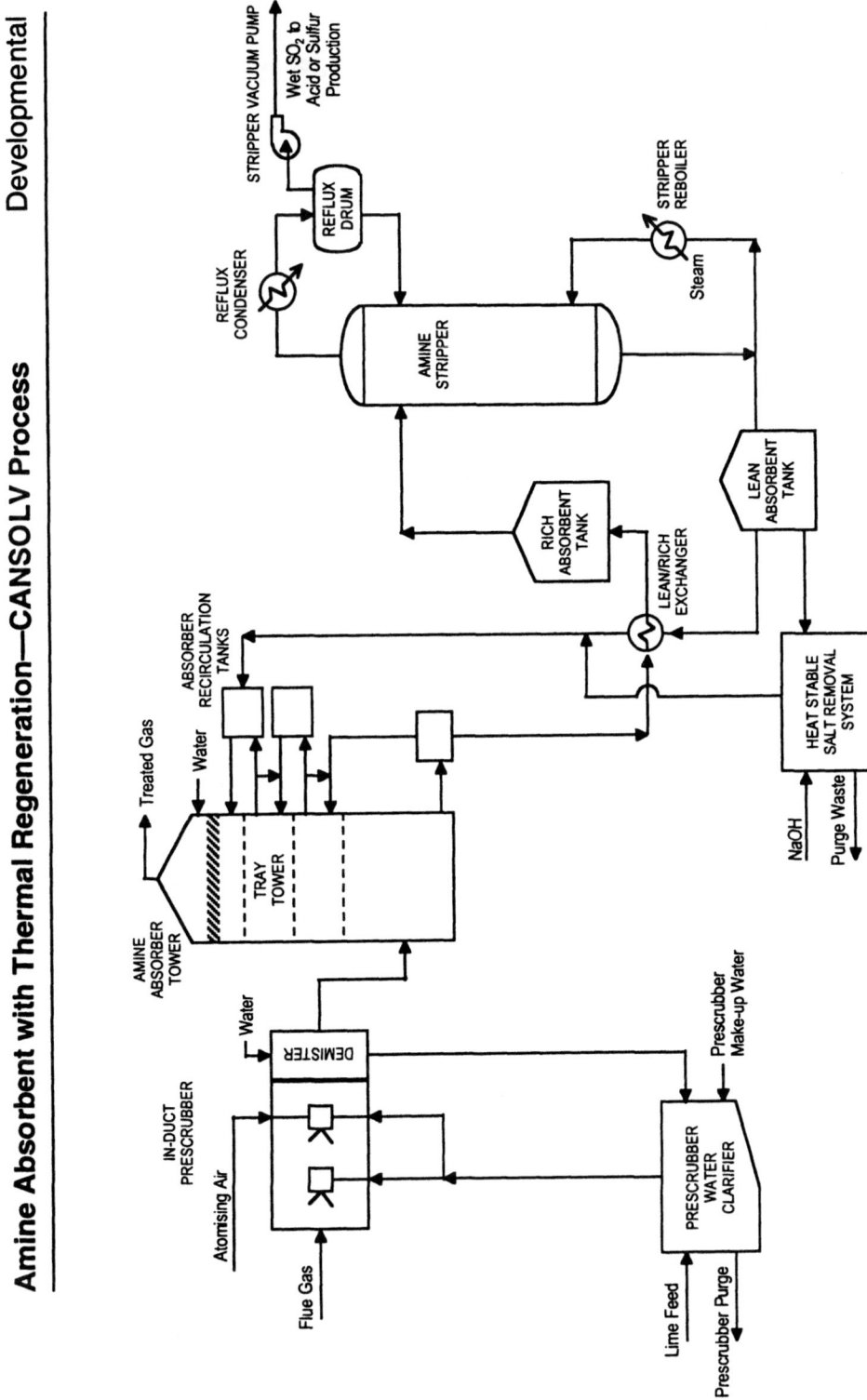

Amine Absorbent with Thermal Regeneration—CANSOLV Process — Developmental

Activated Carbon with Thermal Regeneration

Commercial
Adsorption of SO$_2$ on activated carbon which is thermally regenerated, releasing SO$_2$ for subsequent processing into sulfuric acid or sulfur.

Process Characteristics
SO$_2$ Sorbent: Activated carbon
Principal Raw Materials: Bituminous coal, petroleum coke or other carbon source; Ammonia (for NO$_x$ control); Natural gas for regeneration fuel; Reducing gas (for regeneration associated with conversion to sulfur)
Potentially Saleable Byproducts: SO$_2$-rich gas for conversion to acid or sulfur
Solid Wastes: None
Liquid Effluents: None
Added Gaseous Emissions: None
Inlet SO$_2$ Levels: up to ~3,000 ppm (Commercial Operations)
SO$_2$ Removal: Typical Design Range: 85-98%; Maximum Achievable at Highest SO$_2$: ~95%
NO$_x$ Removal: Typically 60-85%
Particulate Removal: Upstream decoupled particulate control required

Commercial Application
Number/Types: Many: Power plants; refineries; incinerators; metallurgical plants
Locations: Europe; Japan
First Deployment: 1984
Current Status: Active

Process Description
In a typical Mitsui-BF process, flue gas passes through a particulate collector and is quenched by a water spray before entering the activated coke reactor/adsorber. The reactor consists of two compartments, one above the other. Both reactor compartments are filled with granular activated carbon. Flue gas enters the lower reactor compartment where SO$_2$ is adsorbed and catalytically converted to sulfuric acid on the surface of the activated carbon granules. Ammonia is then injected between the two compartments. In the upper section, NO$_x$ in the SO$_2$-free gas is reduced to water vapor and elemental nitrogen by reaction with ammonia at a temperature of 212°-400°F. The carbon slowly moves from the upper to the lower section and then moved by bucket conveyor to the regenerator. In the regenerator, the acid-loaded carbon is first indirectly heated to 550° to 850°F. The adsorbed SO$_3$ is reduced to SO$_2$ which consumes part of the carbon in the process. The regenerated carbon is then indirectly cooled prior to return to the top of the adsorber unit. In the regenerator, any ammonium sulfate formed by oxidation of SO$_2$ and reaction of SO$_3$ with ammonia is reduced to nitrogen, water vapor and SO$_2$. Resulting SO$_2$-rich gas is then passed to further processing to sulfuric acid or elemental sulfur.

Advantages/Disadvantages

Advantages
- Carbon adsorption eliminates handling of solutions and slurries
- Produces SO$_2$–rich byproduct gas
- Integrated SO$_2$ and NO$_x$ removal
- Capable of handling high inlet SO$_2$

Disadvantages/Limitations
- Relatively high capital cost
- SO$_2$ removal limited to about 90% on average
- Need to handle ammonia, a highly regulated hazardous chemical

Principal Suppliers/Developers
Marsulex (Mitsui Licensee)
Mitsui (Mitsui-BF Process)
Steag

Steinmuller/Hugo Peterson
Sumitomo Heavy Industries (EPDC/SHI)

Uhde (Mitsui Licensee)

Byproduct Processes

Carbon Adsorption with Thermal Regeneration — Mitsui-BF Process — Commercial

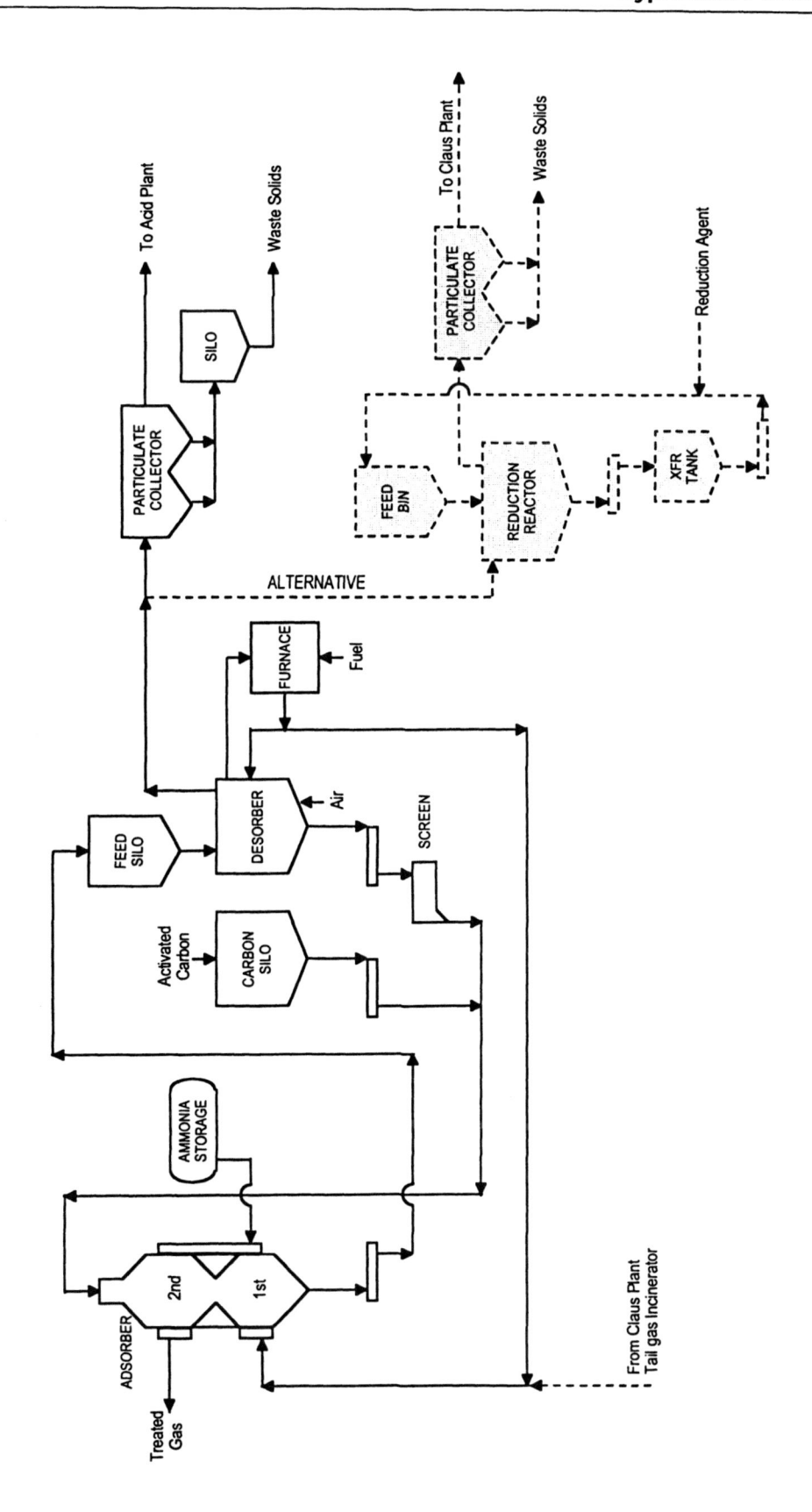

Limestone Clear Liquor Scrubbing

Developmental
Absorption of SO_2 in organic acid buffered solution, then reaction of spent liquor with limestone in a sludge bed reactor to precipitate either gypsum (forced oxidation mode) or waste solids (inhibited oxidation mode).

Process Characteristics
SO_2 Sorbent: Organic acid buffered solution
Principal Raw Materials: Limestone; Organic acid buffer (e.g., dibasic acid, formic acid); Sulfur – Inhibited Oxidation mode only
Potentially Saleable Byproducts: Gypsum (Forced Oxidation mode); CO_2 gas (Inhibited Oxidation mode)
Solid Wastes: Mixed calcium sulfite/sulfate waste solids (Inhibited Oxidation mode)
Liquid Effluents: Purge liquor may be required for commercial grade gypsum
Added Gaseous Emissions: CO_2 gas if discharged (Inhibited Oxidation mode)
Inlet SO_2 Levels: 2,000 ppm (Tested)
SO_2 Removal: 90% (Projected) - but pilot plant equipment has limited removal to <80%
NO_x Removal: Nil – Although additives are under development that can be used for integrated control (e.g., soluble iron chelates)
Particulate Removal: Integrated control capability unless producing commercial grade gypsum

Commercial Application
Number/Types: None
Locations: NA
First Deployment: NA
Current Status: Active (Pilot plant operations continue)

Process Description
Flue gas is contacted in a countercurrent tray tower with an organically buffered aqueous solution which absorbs the SO_2. There are two potential modes of operation being developed, each having a slightly different method of treating the spent solution.

Forced Oxidation mode – Spent solution is first sent to a combination oxidizer/reactor tank where gypsum is precipitated by simultaneous air sparging and limestone addition. Slurry is then fed to a sludge bed reactor tank (SBRT) for crystal growth and concentration. A portion of the SBRT underflow is recycled to the oxidizer/ reactor and the remainder is centrifuged to produce gypsum cake. SBRT overflow is passed to a clarifier/thickener for removal of fines that are recirculated to the SBRT. Clarified liquor is returned to the absorber along with makeup organic acid buffer.

Inhibited Oxidation mode - Spent solution is fed directly to a SBRT where it is reacted with limestone. Carbon dioxide gas is released from the SBRT that can be collected as a potential saleable byproduct. Thickened underflow is centrifuged and overflow is fed to a clarifier/thickener for removal of fines that are recirculated to the SBRT. Organic acid makeup and emulsified sulfur are added to the clarified overflow before return to the scrubber.

Advantages/Disadvantages

Advantages
- Clear solution scrubbing minimizes scaling and plugging problems
- Uses low cost limestone as the reagent
- Capable of integrated SO_2 and particulate control if waste product generated

Disadvantages/Limitations
- For option to produce waste solids may require admixture of the waste with ash (and lime) to make wastes disposable
- Purge liquor with treatment usually necessary for option to produce commercial gypsum

Principal Developers
Electric Power Research Institute and Sponsors

Byproduct Processes

Limestone Clear Liquor Scrubbing

Developmental

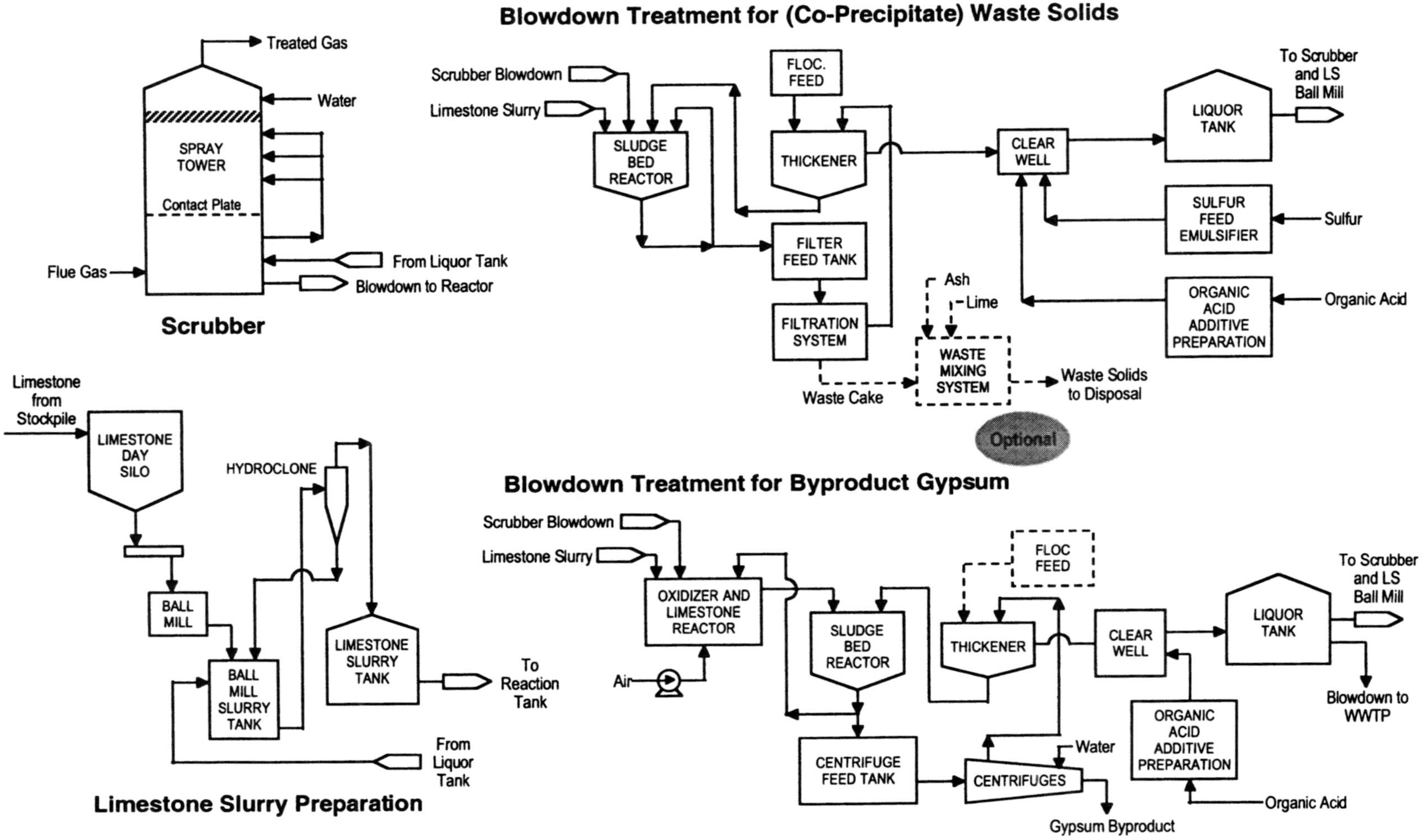

Blowdown Treatment for (Co-Precipitate) Waste Solids

Blowdown Treatment for Byproduct Gypsum

Scrubber

Limestone Slurry Preparation

Kureha Process – Sodium Acetate/Limestone (or Lime) Dual Alkali

Developmental
Flue gas is contacted in a two-part scrubber vessel with both sodium acetate and limestone slurry which absorbs SO_2 forming sulfite salts which are then oxidized to sulfate and reacted with limestone to form gypsum.

Process Characteristics
SO_2 Sorbent: Sodium acetate
Principal Raw Materials: Limestone (or Lime); Acetic acid; Soda ash or caustic
Potentially Saleable Byproducts: Gypsum
Solid Wastes: Blowdown liquor treatment wastes if prescrubber required
Liquid Effluents: Purge liquor may be required for commercial grade gypsum
Added Gaseous Emissions: Potential for loss of a small amount of acetic acid to the stack discharge
Inlet SO_2 Levels: 1,500 to 2,000 ppm (Tested)
SO_2 Removal: 90-95% (Tested)
NO_x Removal: Nil – Although claims that additives (catalysts) can be used to allow simultaneous SO_2 and NO_x control
Particulate Removal: Upstream particulate control required for commercial gypsum

Commercial Application
Number/Types: None
Locations: NA
First Deployment: NA
Current Status: Inactive (Pilot plant work completed)

Process Description
The scrubber vessel consists of two sections, each having multi-stage contacting. In the first section, the gas is contacted with sodium acetate solution where absorption of SO_2 produces sodium sulfite and acetic acid, a portion of which is volatilized. In the second section, the volatilized acetic acid is captured along with additional SO_2 with limestone slurry sprays forming soluble calcium acetate and bisulfite. Spent solutions from both stages are sent to an air-blown oxidizer where the sulfite/bisulfite is converted to sulfate. The acidified sodium sulfate solution is then reacted with limestone to produce gypsum and regenerate sodium acetate. The gypsum is removed by conventional dewatering and the recovered liquor returned to the scrubber. A variation of this process that uses lime rather than limestone has also been pilot tested.

Advantages/Disadvantages

Advantages
- High removal efficiency potential for a wide range of inlet SO_2 levels
- Tolerant of wide fluctuations in inlet SO_2
- Produces commercial grade gypsum
- Tolerant of a wide range of absorbent oxidation

Disadvantages/Limitations
- Use of slurry in the absorber circuit presents possibilities of plugging and scaling
- Moderately complex scrubbing scheme

Principal Developers
Kureha

Kureha Process — Sodium Acetate/Limestone Dual Alkali

Developmental

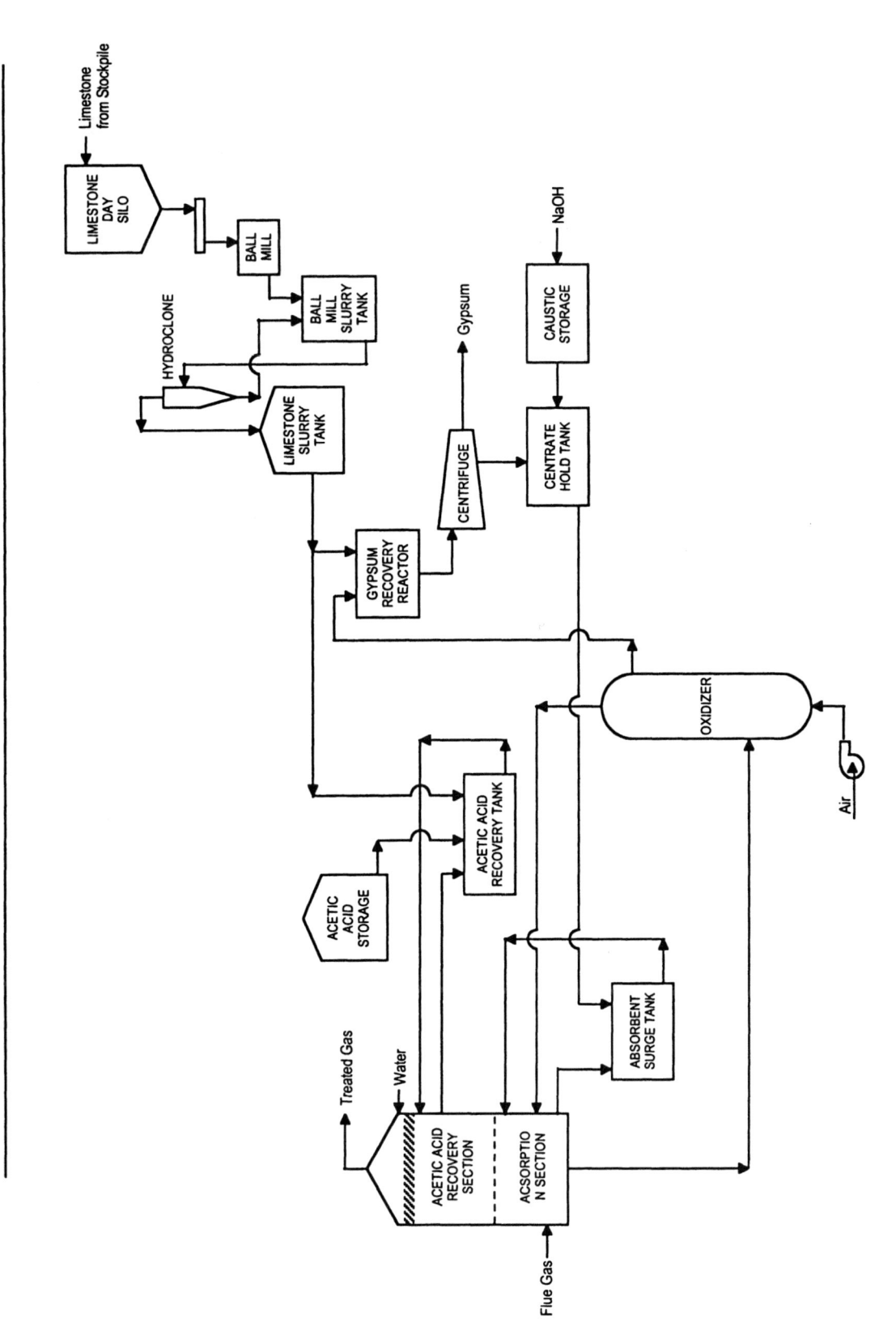

Sodium (Bi)Carbonate Sorption with Ammonia Regeneration

Developmental
Sodium bicarbonate or carbonate is used for dry sorption of SO_2 or a solution of sodium carbonate for SO_2 scrubbing after which spent solution is neutralized and oxidized (or dry solids are dissolved and oxidized). The sodium sulfate is then reacted with ammonia and carbon dioxide to regenerate sodium bicarbonate (or sodium carbonate) for SO_2 sorption and ammonium sulfate fertilizer byproduct.

Process Characteristics
SO_2 Sorbent: Sodium bicarbonate (for dry injection); Sodium carbonate (for dry or solution scrubbing)
Principal Raw Materials: Ammonia; Carbon dioxide; Sodium carbonate
Potentially Saleable Byproducts: Ammonium sulfate fertilizers (Pelletized Sulfur Ammonium Sulfate is currently being pursued)
Solid Wastes: A small amount of precipitated calcium and magnesium carbonate from regeneration system along with precipitated heavy metals; Ash and particulate matter if combined particulate control is employed
Liquid Effluents: None
Added Gaseous Emissions: CO_2 in regeneration system vent gases (in some systems this could be recycled)
Inlet SO_2 Levels: up to ~2,000 ppm (Dry injection); up to ~5,000 ppm (Dry scrubbing); over 100,000 ppm (Solution scrubbing)
SO_2 Removal: Typical Design Range: 50-75% (Dry injection); 80-95% (Dry scrubbing); up to 99%+ (Solution scrubbing)
NO_x Removal: Some NO_x removal can be obtained in dry injection and dry scrubbing system options
Particulate Removal: Combined particulate control is an option

Commercial Application
Number/Types: None
Locations: NA
First Deployment: NA
Current Status: Active (Pilot testing of regeneration system is completed and commercial units are being pursued; scrubbing systems have been commercialized)

Process Description
There are several process variations both in terms of SO_2 scrubbing and absorbent regeneration/ byproduct conversion. The concept centers on the conversion of sodium sulfate to sodium bicarbonate (or carbonate) for SO_2 scrubbing and a premium grade ammonium sulfate byproduct fertilizer. This is accomplished by reacting the sodium sulfate with ammonia and carbon dioxide to produce sodium bicarbonate and ammonium sulfate which can be separated and purified by successive precipitation and crystallization. The sodium bicarbonate can then be calcined to sodium carbonate and the ammonium sulfate pelletized alone or in combination with sulfur to produce high grade Sulfur Ammonium Sulfate (SAS). An option is to convert the ammonium sulfate to potassium sulfate by reaction with potassium chloride and lime which produces a waste $CaCl_2$ brine. The regenerated sodium bicarbonate can be used directly in dry injection SO_2 control, or sodium carbonate can be used in dry scrubbing (spray drying or circulating bed scrubbers) or solution scrubbing.

Advantages/Disadvantages

Advantages
- Produces a premium fertilizer byproduct
- High tolerance for sorbent oxidation
- Wide range in SO_2 removal capability (flexibility in type of scrubbing employed)
- Combined particulate control is an option

Disadvantages/Limitations
- Requires inexpensive supply of ammonia and carbon dioxide
- Relatively complex regenration scheme yet to be demonstrated
- Need to handle ammonia, a highly regulated hazardous chemical

Principal Suppliers
Airborne Technologies - for regeneration system; coupled with any number of suppliers of appropriate scrubbing technologies

Byproduct Processes

Sodium (Bi)Carbonate Sorption with Ammonia Regeneration — Developmental

Dry Scrubbing Alternative Depicted

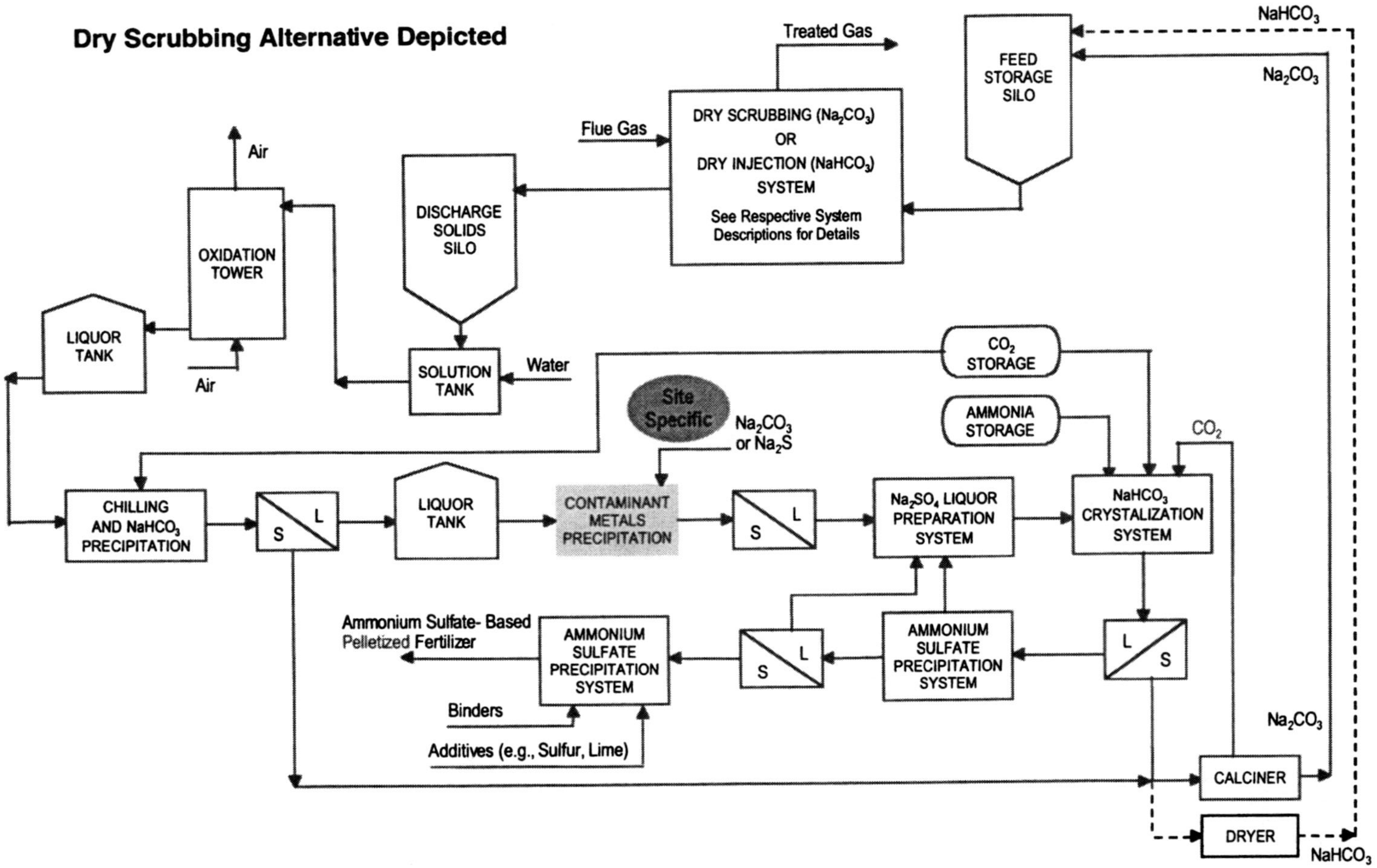

Pircon-Peck Process

Developmental
SO$_2$ is absorbed in a slurry of calcium and ammonium pyrophosphates with neutralization of spent slurry using ammonia and then partially dewatering the solids to produce a fertilizer byproduct slurry.

Process Characteristics
SO$_2$ Sorbent: Ammonium and calcium pyrophosphates
Principal Raw Materials: Ammonia; Phosphate rock; Phosphoric acid
Potentially Saleable Byproducts: Fertilizer "slurry" mix of ammonium pyrophosphate (solution) with calcium pyrophosphate, calcium sulfite and calcium sulfate (solids)
Solid Wastes: None
Liquid Effluents: None
Added Gaseous Emissions: None
Inlet SO$_2$ Levels: 1,500 to 5,000 ppm (Tested)
SO$_2$ Removal: 90-95% (Tested)
NO$_x$ Removal: Nil
Particulate Removal: Integrated control capability (but only if "product" marketability is not adversely affected)

Commercial Application
Number/Types: None
Locations: NA
First Deployment: None
Current Status: Inactive

Process Description
The process involves the use of a scrubbing liquor containing a mix of ammonium pyrophosphates (solution) and calcium pyrophosphates (solid) in water. Flue gas is contacted in two absorption towers in series and generally in countercurrent flow the regenerated absorbent. The first tower humidifies the gas, concentrates the absorbent liquor and removes most of the SO$_2$. The second tower serves as the final SO$_2$ cleanup step. Spent absorbent slurry is fed to a two-stage reactor system where calcium pyrophosphate, ammonia and makeup are added. Ammonia controls the absorbent solution pH and calcium pyrophosphate precipitates calcium sulfite and sulfate. The regenerated slurry is then passed to a thickener where a portion of the slurry is thickened as a byproduct fertilizer slurry and the remainder recycled to the absorber towers. It would be preferable to market the fertilizer byproduct in slurry form to minimize the processing costs and allow greater flexibility in the ratios of nitrogen and phosphorous to calcium and sulfur. Integrated particulate control would add a fly ash constituent that could enhance the value as a soil amendment or conditioner.

Advantages/Disadvantages

Advantages
- Produces potentially marketable fertilizer slurry byproduct
- Capability for integrated SO$_2$ and particulate control
- High tolerance for sorbent oxidation
- Phosphate chemistry precludes potentia for a "blue plume" in the stack

Disadvantages/Limitations
- Process economic feasibility hinge importantly on the availability, form and cost of the raw materials and marketability of the byproduct
- Potential for high sorbent attrition losses
- Need to handle ammonia, a highly regulated hazardous chemical

Principal Developers
Illinois Institute of Technology and Resources Agricultural Management

Pircon-Peck Process

Developmental

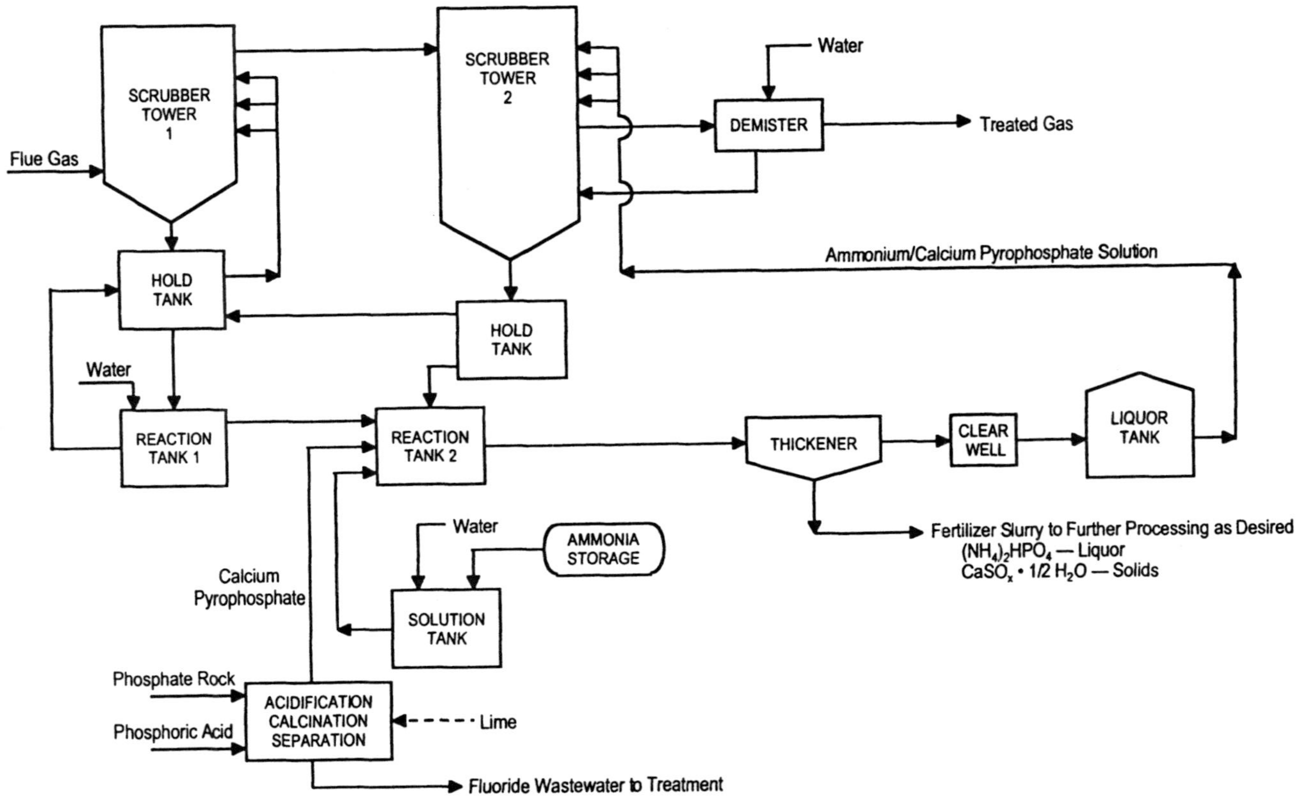

Passamaquoddy Recovery Process

Developmental
SO$_2$ and CO$_2$ are absorbed in a slurry of calcium and potassium oxide producing CaCO$_3$ and potassium sulfate which are recovered as, respectively, kiln feed and fertilizer byproduct.

Process Characteristics
SO$_2$ Sorbent: Mixed Calcium and potassium (and sodium) oxide
Principal Raw Materials: Waste kiln dust
Potentially Saleable Byproducts: Potassium sulfate (fertilizer additive); distilled water; kiln feedstock (CaCO$_3$)
Solid Wastes: None
Liquid Effluents: None
Added Gaseous Emissions: None
Inlet SO$_2$ Levels: 100 to 1,800 ppm (Tested)
SO$_2$ Removal: 92% (Tested)
NO$_x$ Removal: Nil
Particulate Removal: None

Commercial Application
Number/Types: None
Locations: NA
First Deployment: None (DOE Clean coal demonstration project completed in 1994)
Current Status: Inactive

Process Description
Hot cement kiln exhaust gas is first passed through an electrostatic precipitator where entrained cement kiln dust is removed. This waste kiln dust is a mix of CaO and alkali metal oxides (mainly potassium), so it has no value for recycle as a cement kiln feedstock. The recovered dust is stored in a silo for use as the alkali sorbent for the scrubber. The gas is then cooled in a gas-liquid heat exchanger that serves as the principal heat source for the byproduct crystallizer. Cool gas is then contacted with a 20-25% slurry of the cement kiln dust in a gas bubbling reactor. The kiln dust slurry is prepared with fresh and recycled clarified process liquor. Several reactions take place in the reactor from absorption of both SO$_2$ and CO$_2$. CaO is converted to CaCO$_3$ producing a slurry in a solution of potassium sulfate. Spent slurry is fed to a settling tank from which clarified solution is fed to the crystallizer where potassium sulfate solids are precipitated. A bleed stream is of potassium sulfate slurry is sent to a centrifuge to dewater the potassium sulfate solids. The crystallizer also produces distilled water. A portion of the distilled water is mixed with the underflow from the settling tank to dilute the solids before being dewatered in a second settling tank. Dilution of the underflow prior to secondary dewatering renders the recovered CaCO$_3$ solids acceptable as a kiln feedstock. Clarified liquor from the second settling tank is recycled for use in kiln dust slurry preparation.

Advantages/Disadvantages

Advantages
- Produces potentially marketable fertilizer byproduct
- Does not require a purchased sorbent raw material
- Recovers cement feedstock from a kiln waste product

Disadvantages/Limitations
- Only applicable to cement kilns or other plants with access to kiln dust
- Impure potassium sulfate byproduct may have limited market value
- Complicated scrubber design with high plugging and scale potentia

Principal Developers
Passamaquoddy Technology

Byproduct Processes

Passamaquoddy Recovery Process — Commercial

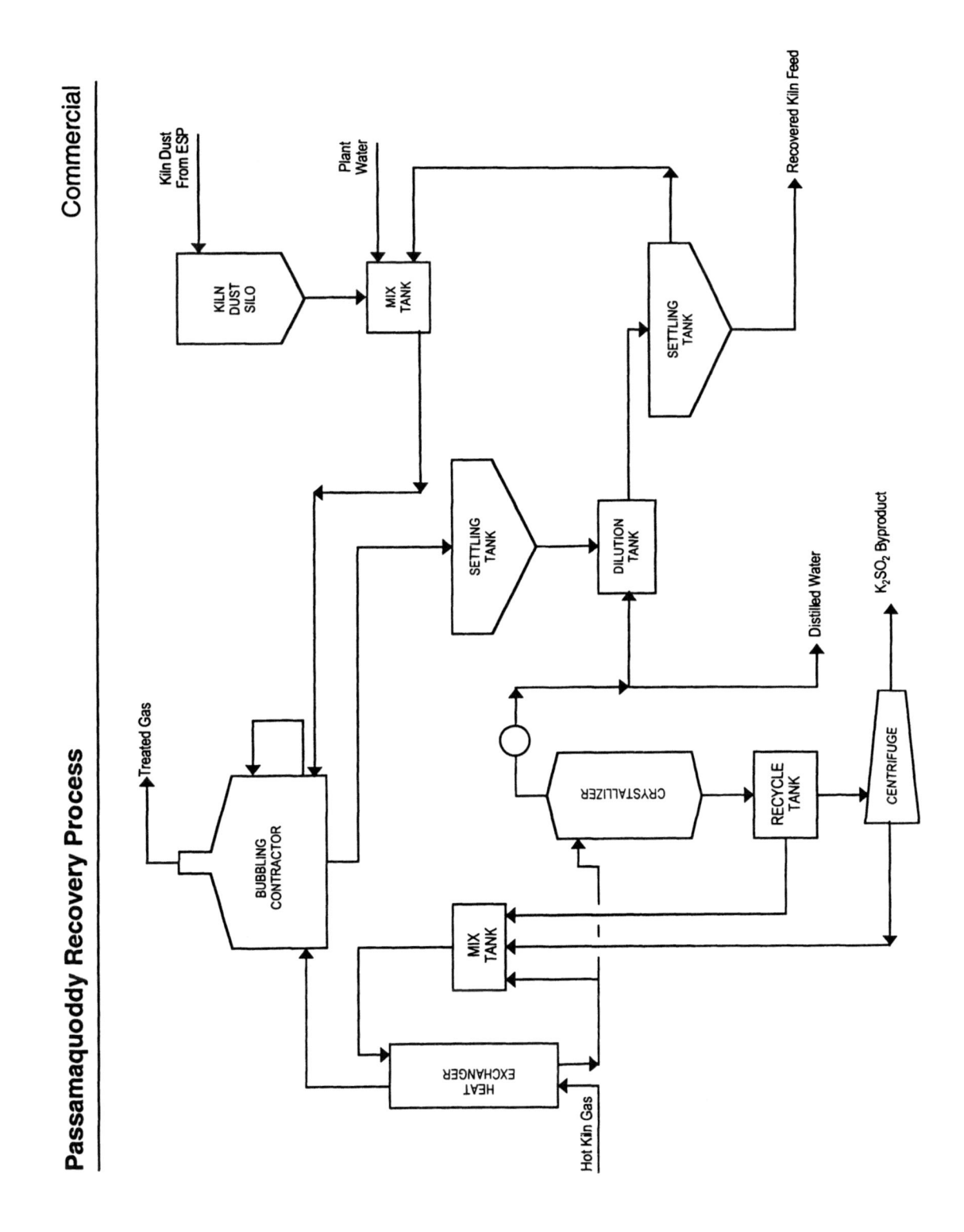

ISPRA Process

Developmental
Absorption and oxidation of SO_2 in an acidified bromine solution to form sulfuric and hydrobromic acids. Sulfuric acid is then concentrated to 95% and bromic acid is electrolytically regenerated to bromine for reuse.

Process Characteristics
SO_2 Sorbent: Acidic bromine solution (15 wt% H_2SO_4, 15 wt% HBr, 0.5 wt% Br_2)
Principal Raw Materials: Bromine
Potentially Saleable Byproducts: Sulfuric acid (95%); Hydrogen
Solid Wastes: None
Liquid Effluents: Purge liquor for soluble impurities buildup
Added Gaseous Emissions: Hydrogen (from electrolytic cells); Potential for loss of a small amount of bromine to the stack gas discharge
Inlet SO_2 Levels: up to 3,000 ppm (Tested)
SO_2 Removal: 95+% (Claimed)
NO_x Removal: Nil
Particulate Removal: Upstream decoupled particulate control is required

Commercial Application
Number/Types: None
Locations: NA
First Deployment: NA
Current Status: Unknown (~30,000 cfm Pilot Plant Operation)

Process Description
Hot flue gas is first utilized to provide some of the thermal load for sulfuric acid concentration and treated gas reheat. Prior to contact for SO_2 removal, the flue gas is cooled in these evaporators and is further cooled in gas-to-gas regenerative heat exchanger. The cooled flue gas is then passed to the Reactor where it is contacted with acidified bromine solution, absorbing and oxidizing the SO_2 to sulfuric acid and hydrobromic acid. After a demisting tower, the desulfurized gas passes through a final scrubbing stage as a final cleaning step to remove residual bromine solution. Solution from the Reactor is continuously withdrawn and a portion sent to the pre-concentrator and a portion to an electrolytic regeneration system. In the electrolytic cells, hydrogen is evolved at the cathode and bromine at the anode. The bromine is returned to the process and the hydrogen is purified as a byproduct. Testing conducted to date reportedly show that the presence of chlorides and NO_x in the flue gas does not affect the performance of the system, although there is no definitive discussion of how these impurities are managed.

Advantages/Disadvantages

Advantages
- Clear solution scrubbing minimizes scaling and plugging problems
- Produces byproduct concentrated H_2SO_4

Disadvantages/Limitations
- Bromine is considered hazardous materials and is a highly regulated toxic material
- Full scale operating parameters have not been demonstrated in integrated operations

Principal Developers
European Community's Institute of Environmental Sciences
Ferlini/General Atomics

ISPRA Process

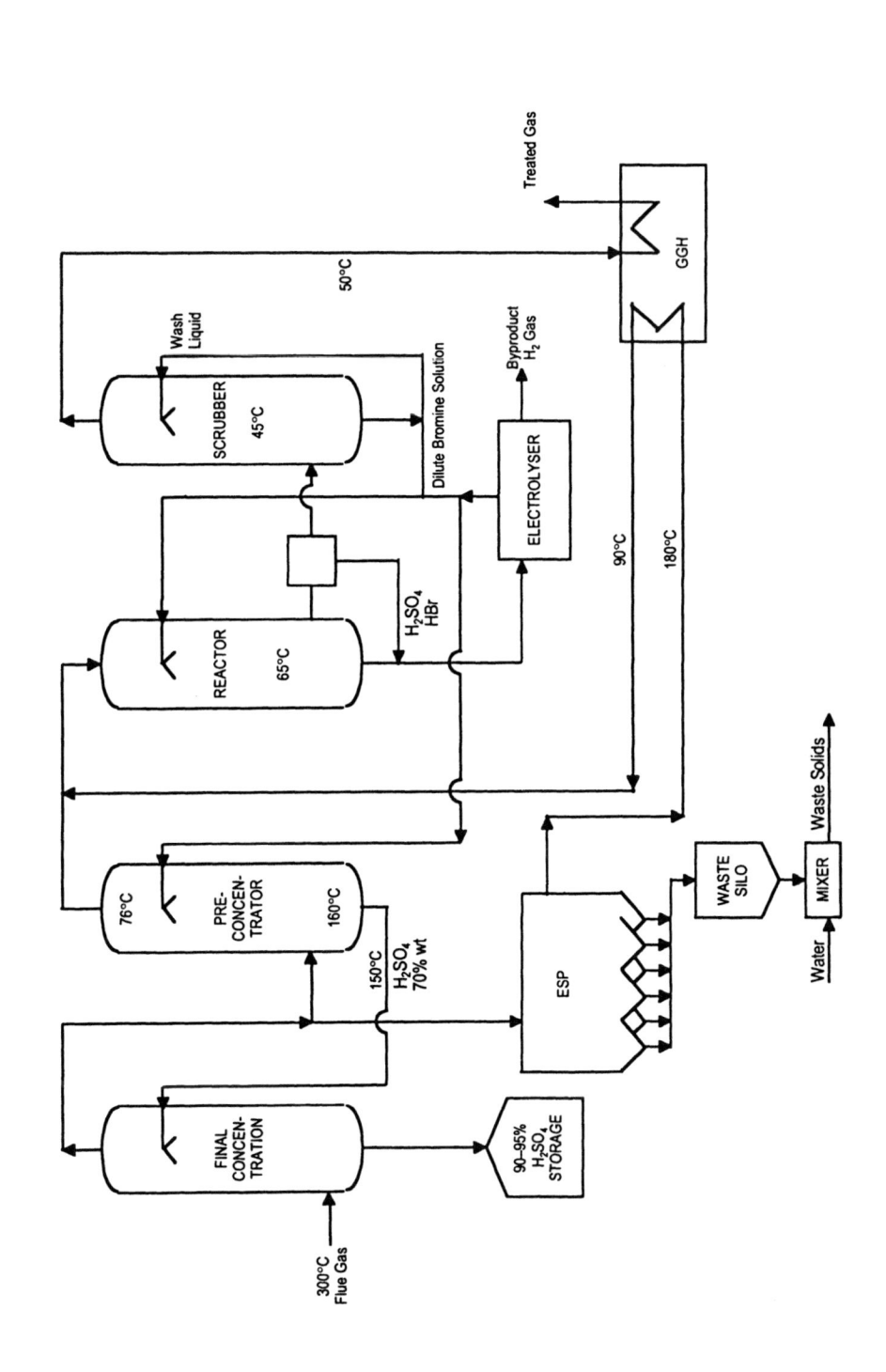

Developmental

Electrochemical Membrane Separation

Developmental
SO$_2$ from hot flue gas at about 750°F is collected through flow-by electrodes and oxidized after which it migrates through membranes to an anode, is converted to SO$_3$ and collected as oleum in a storage tank.

Process Characteristics
SO$_2$ Sorbent: Electrochemical separation
Principal Raw Materials: Nitrogen
Potentially Saleable Byproducts: Oleum
Solid Wastes: None
Liquid Effluents: None
Added Gaseous Emissions: None
Inlet SO$_2$ Levels: <1,000 to ~5,000 ppm (Tested)
SO$_2$ Removal: 99% Capability (Claimed)
NO$_x$ Removal: Presently being developed
Particulate Removal: None

Commercial Application
Number/Types: None
Locations: NA
First Deployment: NA
Current Status: Active (Currently in small scale pilot testing)

Process Description
The technology involves a series of "sandwich" style cells consisting of a cathode, an electrolyte membrane and an anode. Hot flue gas at ~750°F (e.g., upstream of a boiler air preheater) contacts flow-by porous cathodes that collect and oxidize the SO$_2$ to SO$_3$. (In a variation of the process, SO$_2$ is catalytically converted to SO$_3$ in a catalytic reactor similar to that used in direct sulfuric acid processes described elsewhere.) Sulfate ions then migrate through electrolyte membrane under the influence of a low-voltage applied potential to the anode. A nitrogen sweep recycle gas then picks up the SO$_3$ and carries it through a recovery unit such as that used in a sulfuric acid plant to produce oleum (SO$_3$ in H$_2$SO$_4$). The electrodes are constructed of a crystalline ceramic material; the membrane electrolyte is an immobilized molten salt; and the cell housings are glass-ceramic. Ongoing development work is focusing on NO$_x$ removal by reduction to N$_2$, optimizing the cell matrix material and improved membrane design to decrease polarization and improve current densities.

Advantages/Disadvantages

Advantages
- No chemicals or reagents other than nitrogen makeup
- Produces oleum byproduct - no waste products
- Low operating labor potential
- Claimed capability of ~85% SO$_3$ collection
- Potential for simultaneous NO$_x$ control
- Stack gas requires no reheat

Disadvantages/Limitations
- High capital investment associated with hot gas processing and specialty materials of construction
- Reliability issues relating to the types of materials being developed not thoroughly tested in FGD-style commercial applications

Principal Developers
Georgia Institute of Technology

Electrochemical Membrane Process

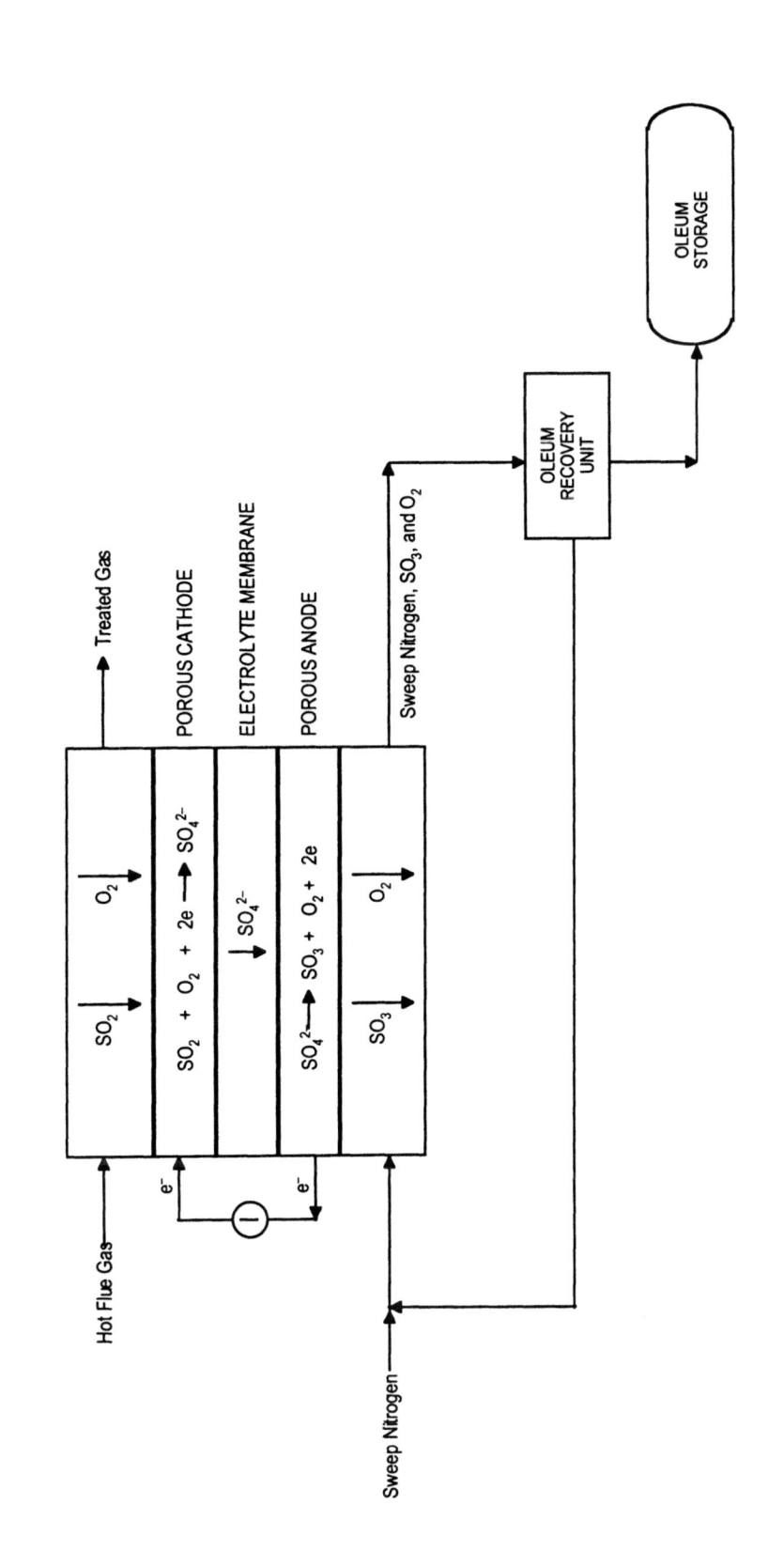

Developmental

$SO_2 + O_2 + 2e \rightarrow SO_4^{2-}$

$SO_4^{2-} \rightarrow SO_3 + O_2 + 2e$

Sulfuric Acid Absorption with Peroxide Oxidation

Developmental
HCl and SO_2 are removed from combustion gas, first HCl in concentrated HCl, then SO_2 in concentrated sulfuric acid that is cycled through an electrolytic cell to produce peroxide to enhance absorbed SO_2 oxidation.

Process Characteristics
SO₂ Sorbent: Sulfuric acid (for SO_2); Hydrochloric acid (for HCl)
Principal Raw Materials: (Lime for treatment of prescrubber blowdown treatment)
Potentially Saleable Byproducts: Sulfuric acid (~95%); Hydrochloric acid (~31%)
Solid Wastes: Prescrubber waste liquor blowdown treatment
Liquid Effluents: Prescrubber treated liquor
Added Gaseous Emissions: Hydrogen from the electrolytic cells for peroxide production
Inlet SO₂ Levels: 500 to 2,000 ppm (Tested)
SO₂ Removal: ~90%(Tested)
NOₓ Removal: Nil
Particulate Removal: Upstream decoupled particulate control required

Commercial Application
Number/Types: None
Locations: NA
First Deployment: None
Current Status: Inactive

Process Description
The process is primarily focused on treating flue gases with both HCl and SO_2 removal requirements such as from municipal solid waste (MSW) incinerators. Gas is first prescrubbed to humidify the gas and remove non-acidic contaminates (e.g., residual particulate, heavy metals). The prescrubber is operated at low pH levels to minimize absorption of HCl and SO_x. The gas then passes to a two-stage hydrochloric acid absorber. Spent acid solution from the absorber is sent to an acid distillation concentrator that produces a 31% strength acid product and a lean acid solution recycled to the absorber. Downstream of the HCl absorber, the gas passes through a high efficiency demister to remove HCl solution droplets. It then enters the SO_2 absorber where it is contacted with concentrated sulfuric acid which absorbs both SO_2 and SO_3. Air is also blown into the absorber to effect a first stage of SO_2 to SO_3 oxidation. The acid solution is continuously recirculated through an electrolytic cell operating in a pure acid persulfate mode which produces peroxide in situ which completes oxidation of residual SO_2. Product sulfuric acid at 95% strength is withdrawn from the electrolyic cell discharge.

Advantages/Disadvantages

Advantages
- Produces two byproduct acids of commercial strength

Disadvantages/Limitations
- Requires high efficiency particulate control upstream
- Economically attractive with gases containing both significant levels of HCl and SO_2
- Requires a prescrubber which produces a purge liquor requiring treatment

Principal Developers
Noell/KCR

Sulfuric Acid Absorption with Peroxide Oxidation—Noell/KRC Process   Developmental

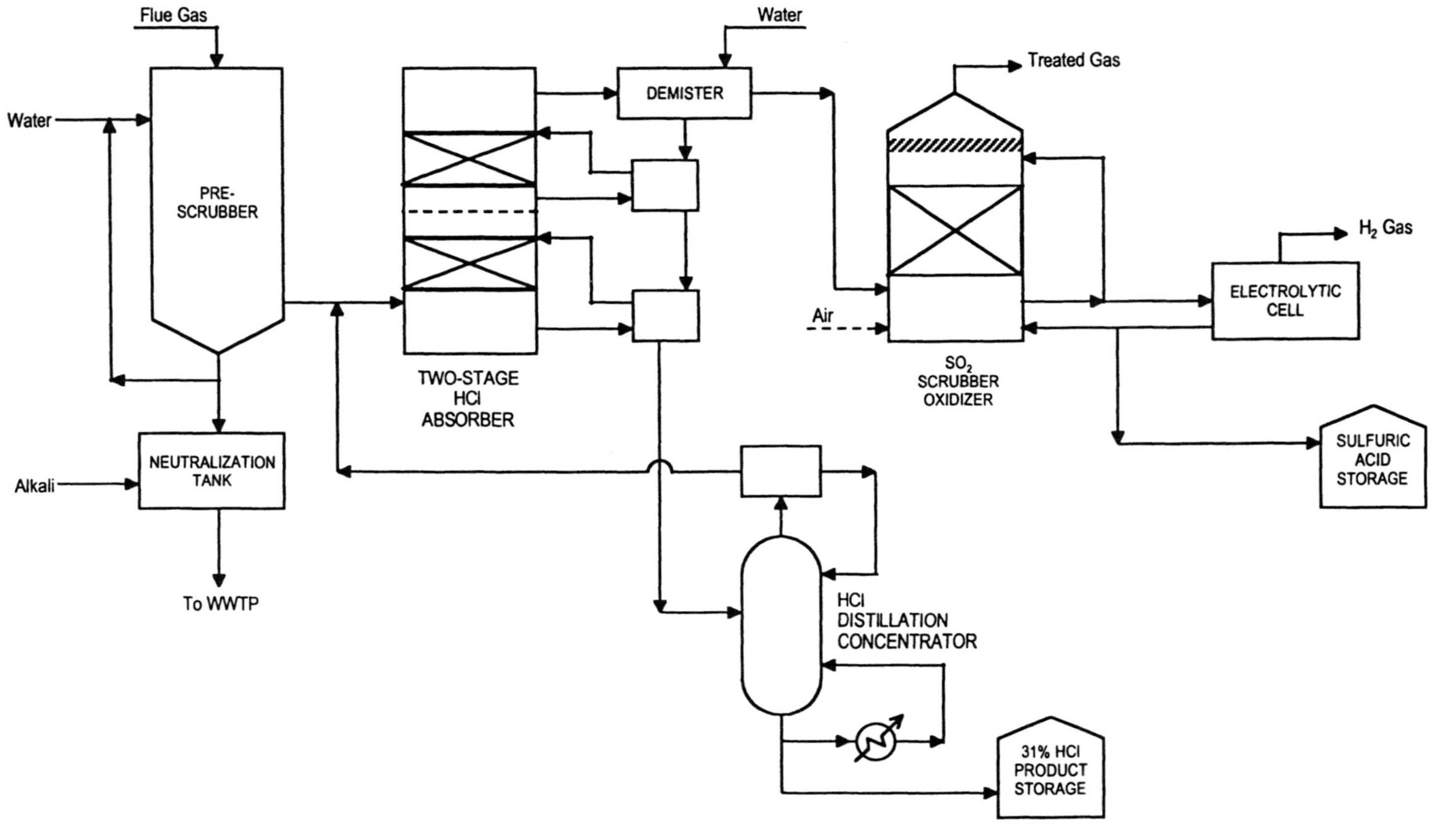

Carbon Adsorption with Acid Regeneration

Developmental
Downstream of a high efficiency particulate collector, SO_2 is adsorbed on a activated carbon which is then regenerated with a succession of sulfuric acid solutions that is finally concentrated to 65% strength.

Process Characteristics
SO_2 Sorbent: Activated carbon
Principal Raw Materials: Activated carbon
Potentially Saleable Byproducts: Sulfuric acid (~65%)
Solid Wastes: None
Liquid Effluents: None
Added Gaseous Emissions: None
Inlet SO_2 Levels: up to ~2,000+ ppm (Tested)
SO_2 Removal: ~80% Average(Tested)
NO_x Removal: Nil
Particulate Removal: Upstream decoupled particulate control required

Commercial Application
Number/Types: None
Locations: NA
First Deployment: None
Current Status: Inactive

Process Description
Flue gas from an upstream high efficiency, dry particulate collector (e.g., fabric filter or ESP) is contacted in a set of parallel activated carbon adsorber units. SO_2 is adsorbed on the activated carbon. The adsorber units are configured to position most of the units in the adsorption mode, with isolation of one or more units for regeneration. As units are cycled through regeneration they are washed with successively more dilute solutions of sulfuric acid, starting with about 20% strength and ending with water. Each stage of regeneration wash is segregated and "indexed" forward. The first wash of most concentrated acid at 20% is discharged to a submerged combustor evaporator to fully oxidize the absorbed SO_2 and concentrate the acid solution to about 65%. The concentrated acid is then filtered to remove particulate matter and stored for product distribution. The process has been operated in several large-scale demonstration plants. The last, a 55 MWe system, was operated for over five years. But, the process has never been commercialized, due in part to the limited market for the concentration of acid produced.

Advantages/Disadvantages

Advantages
- Produces moderately concentrated sulfuric acid (~65%) directly
- No raw materials other than replacement carbon
- Stack gas requires no reheat

Disadvantages/Limitations
- Requires high efficiency particulate control upstream
- Low acid strength may limit use

Principal Developers
Hitachi
Research Triangle Institute

Byproduct Processes

Carbon Adsorption with Acid Regeneration—Hitachi Process Developmental

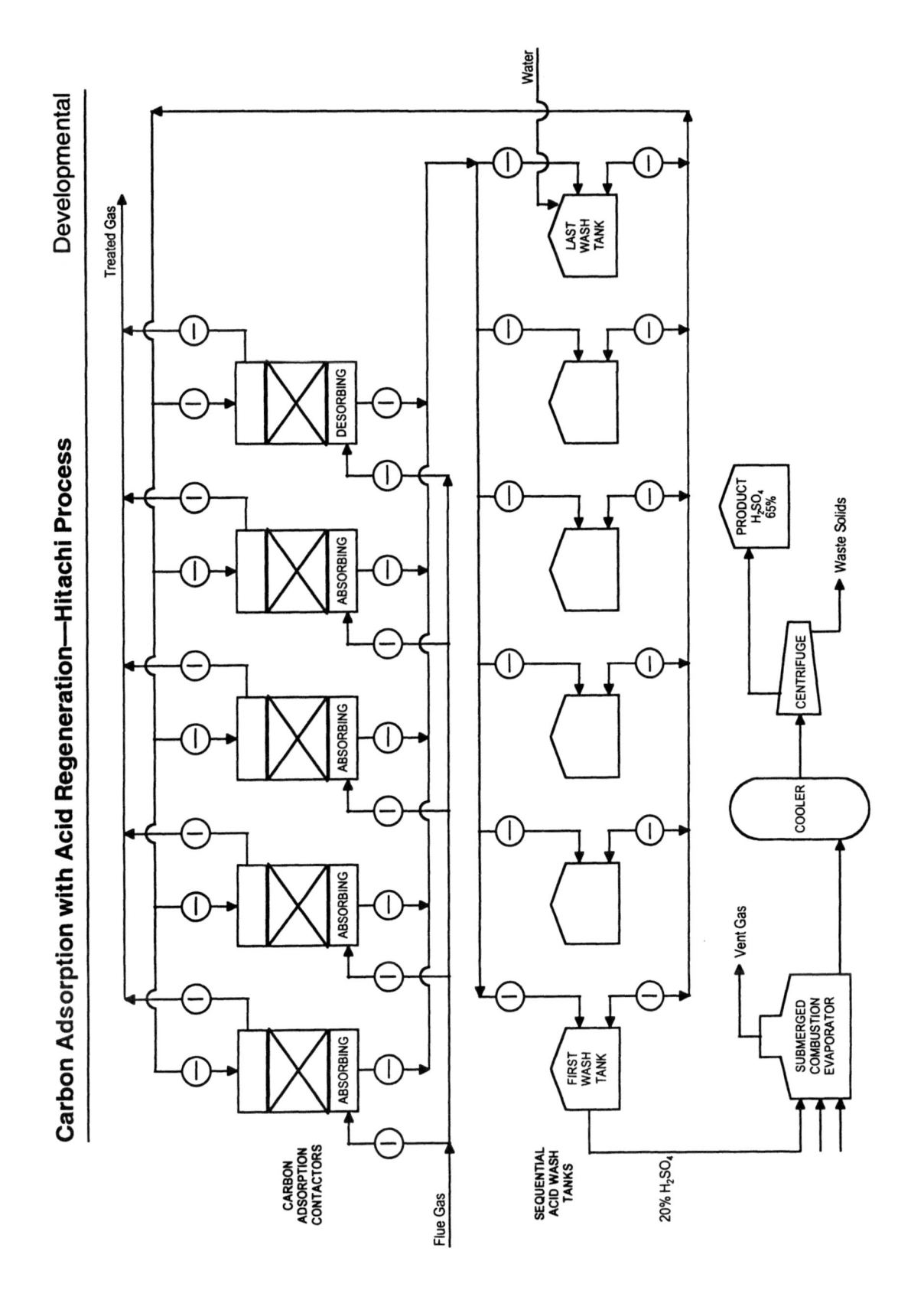

114   Profiles in Flue Gas Desulfurization

Zinc Oxide Process (Direct Slurry Sorption) with NO$_x$ Control

Developmental
Absorption of SO$_2$ and NO$_x$ in a zinc oxide slurry utilizing a spray dryer, then calcination of the solids to evolve gas rich in SO$_2$ and NO$_x$ for further processing to sulfuric and nitric acids.

Process Characteristics
SO$_2$ Sorbent: Zinc oxide/hydroxide
Principal Raw Materials: Zinc Oxide
Potentially Saleable Byproducts: SO$_2$-rich and NO$_x$-rich gas for conversion to sulfuric and nitric acids
Solid Wastes: Purge required for gases containing impurities (e.g., chlorides)
Liquid Effluents: None
Added Gaseous Emissions: None
Inlet SO$_2$ Levels: Not defined
SO$_2$ Removal: Not defined
NO$_x$ Removal: Nil
Particulate Removal: Upstream decoupled particulate removal required

Commercial Application
Number/Types: None
Locations: NA
First Deployment: NA
Current Status: Active (In pilot plant stage)

Process Description
Flue gas is contacted in a spray dryer with a slurry of zinc oxide which both absorbs both SO$_2$ and NO$_x$. Absorption reactions result in a mixture of solids consisting of zinc oxide (unreacted), zinc sulfite, zinc sulfate and a complex mix of hydroxylamine sulphonates. As in any other spray dryer process, the solids are collected in a downstream high efficiency particulate collector (fabric filter or electrostatic precipitator). The collected solids are then calcined under slightly reducing conditions which releases SO$_2$, NO$_x$ and water vapor. The gas is sent to a multistage acid plant for producing sulfuric and nitric acid and the regenerated zinc oxide solids are recycled to the spray dryer feed. The exact configuration of the acid plant has yet to be fully defined. Also the need for solids purge to control buildup of impurities has not been determined.

Advantages/Disadvantages

Advantages
- Use of a spray dryer greatly simplifies process configuration
- Combined SO$_2$ and NO$_x$ control
- Produces byproduct sulfuric and nitric acids
- Stack gas requires no reheat

Disadvantages/Limitations
- Use of a spray dryer type contactor limits applications to inlet SO$_2$ levels to about 3,000 ppm with associated removal efficiencies of <~95%
- Attrition and purge losses may be significant

Principal Developers
Battelle

Byproduct Processes

Zinc Oxide Process (Direct Slurry Sorption) with NO_x Control

Developmental

ELSORB Process

Developmental
Absorption of SO$_2$ a solution of sodium phosphate from which absorbed SO$_2$ is thermally stripped which regenerates the absorbent and produces an SO$_2$-rich gas for further processing.

Process Characteristics
SO$_2$ Sorbent: Sodium phosphate solution
Principal Raw Materials: Caustic or soda ash; Lime (for treating prescrubber blowdown)
Potentially Saleable Byproducts: SO$_2$-rich gas for conversion to sulfuric acid or sulfur; Sodium sulfate
Solid Wastes: Wastes from the prescrubber blowdown treatment
Liquid Effluents: Purge liquors required to control soluble impurities buildup
Added Gaseous Emissions: None
Inlet SO$_2$ Levels: Not Defined
SO$_2$ Removal: Not Defined
NO$_x$ Removal: Nil
Particulate Removal: Upstream decoupled particulate control is required

Commercial Application
Number/Types: None
Locations: NA
First Deployment: NA
Current Status: Inactive

Process Description
The ELSORB process is an analog of the Wellman Lord process except that it uses sodium phosphate rather than sodium sulfite as the absorbent. Flue gas is first passed through a prescrubber to remove particulate and soluble impurities that are discharged through a prescrubber blowdown treatment system. The humidified gas is then contacted with a solution of sodium phosphate which absorbs SO$_2$. The solution is reported to be more highly buffered than the sulfite/bisulfite solution used in the Wellman Lord system, so the flow rates are lower and the equipment somewhat smaller. Most of the spent absorbent liquor is fed to an evaporation system where SO$_2$ is stripped off and sodium phosphate absorbent solution is regenerated. Unlike the Wellman Lord process, there is very little crystal formation in the evaporators so essentially no solids in the circulation loop and no encrustation in the evaporators. The SO$_2$-rich gas is then cooled to remove water vapor prior to being sent to an acid or Claus plant. A bleed stream of spent absorbent is sent to a purge crystallizer to remove sodium sulfate formed from oxidation of absorbed SO$_2$. A small purge stream is also bled from the system to control build up of impurities such as chlorides.

Advantages/Disadvantages

Advantages
- Clear solution scrubbing should minimize scaling in the absorber
- Applicable to a wide range of inlet SO$_2$
- Produces SO$_2$-rich gas for further processing

Disadvantages/Limitations
- Requires a purge stream to control impurities
- A purge stream containing phosphate may be more difficult to deal with from an environmental perspective

Principal Developers
Elkem Technologies

Byproduct Processes

ELSORB Process

Developmental

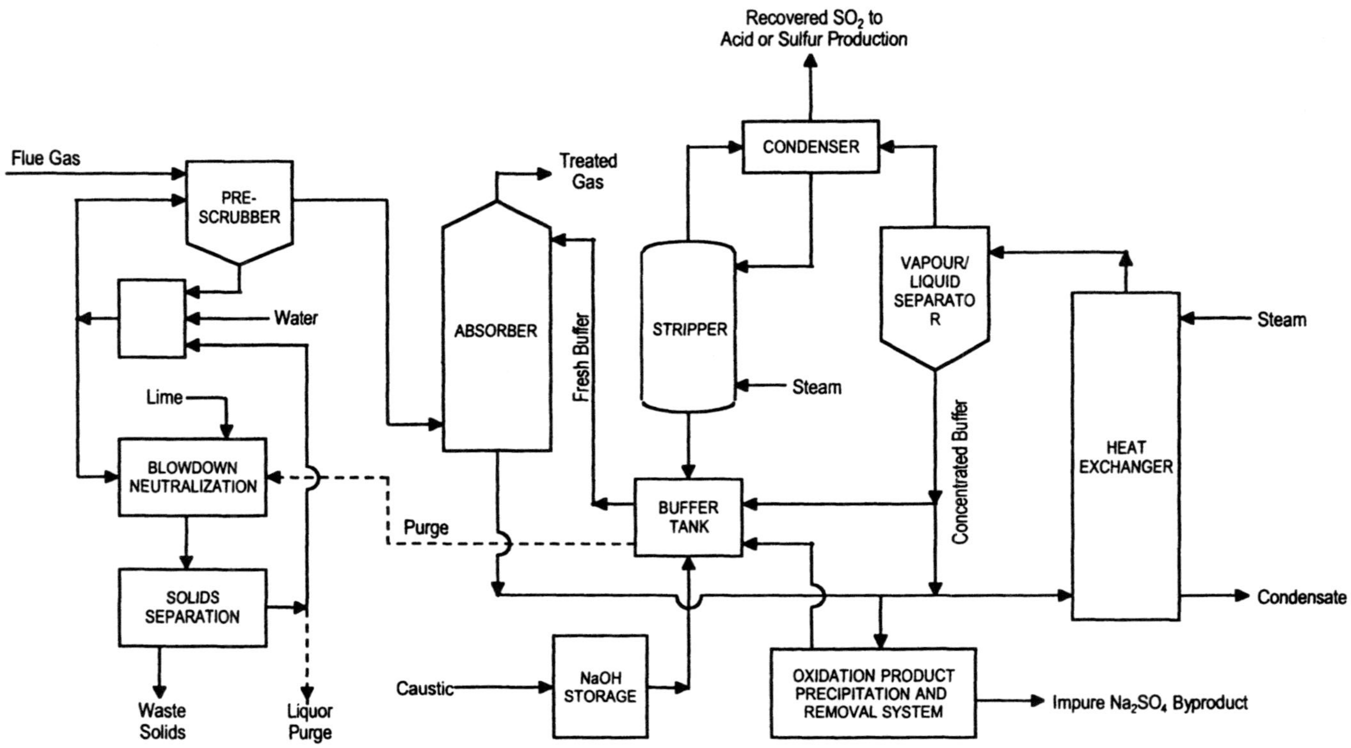

Ammonia Scrubbing with Thermal Regeneration

Developmental
SO$_2$ is absorbed in a solution of ammonium sulfate/sulfite followed by thermal stripping of the spent solution to release SO$_2$ for conversion to sulfur or sulfuric acid.

Process Characteristics
SO$_2$ Sorbent: Ammonia
Principal Raw Materials: Ammonia; (Sulfur, in the IFP process for conversion to SO$_2$-rich gas); (Reducing gas, in the IFP process for conversion to sulfur)
Potentially Saleable Byproducts: SO$_2$-rich gas for conversion to sulfur or acid; alternate for direct sulfur; (Some byproduct solid ammonium sulfate in the TVA process)
Solid Wastes: None
Liquid Effluents: None
Added Gaseous Emissions: Potential for "blue plume" in stack gas discharge; Incinerated sulfur converter vent gas if not recycled to scrubber
Inlet SO$_2$ Levels: up to ~5,000 ppm (Tested)
SO$_2$ Removal: 90+% (Tested)
NO$_x$ Removal: Technology has been proposed for the IFP process to incorporate NO$_x$ control as an integral part of the technology
Particulate Removal: Upstream decoupled, dry particulate control required

Commercial Application
Number/Types: None – A "commercial" system was briefly operated on a copper smelter in Eastern Europe, but shut down
Locations: NA
First Deployment: NA
Current Status: Inactive (completed pilot plant testing)

Process Description
There have been numerous develop programs for ammonia-based absorption/desorption technology for SO$_2$ control. None have achieved commercialization; but, one of the most advanced in terms of development is the joint IFP/Catalytic process with the adjunct CEC (Chisso Engineering Co.) process for combined NO$_x$ control. The process uses conventional multi-stage scrubbing a solution of ammonium sulfite, bisulfite and sulfate (the latter from oxidation of absorbed SO$_2$). There are two approaches to treating the spent liquor. One produces SO$_2$-rich gas for conversion to sulfuric acid or elemental sulfur in an onsite unit; the other produces sulfur directly. In both, the spent liquor is first sent to a countercurrent evaporator where the combination of internal acidification and heat release SO$_2$ and produce an ammonium sulfate slurry. The slurry is then acidified with recycled ammonium bisulfate, then heated in a molten salt bath to drive off ammonia and produce ammonium bisulfate. A small amount of sulfur is also added to the bath to reduce some of the bisulfate to SO$_2$, ammonia and water. Most of the ammonia is condensed and both the gas and ammonia are returned to the scrubber. A similar process without the use of sulfur was developed by TVA using crystallization of ammonium sulfate prior to decomposition. A small byproduct stream of ammonium sulfate accounted for formation.

Advantages/Disadvantages

Advantages
- Produces SO$_2$-rich gas for conversion to sulfuric acid or elemental sulfur
- No solid wastes or liquid effluents
- High SO$_2$ removal capabilities at high inlet SO$_2$

Disadvantages/Limitations
- Upstream particulate removal is required
- Potential for "blue plume" opacity issues requires very high mist removal or a wet ESP
- Need to handle ammonia, a highly regulated hazardous chemical

Principal Developers
Cominco (Exorption Process)   IFP/Catalytic/CEC (Stackpol 150)   TVA (ABS Process)

Byproduct Processes

Ammonia Scrubbing with Thermal Regeneration — IFP Process

Developmental

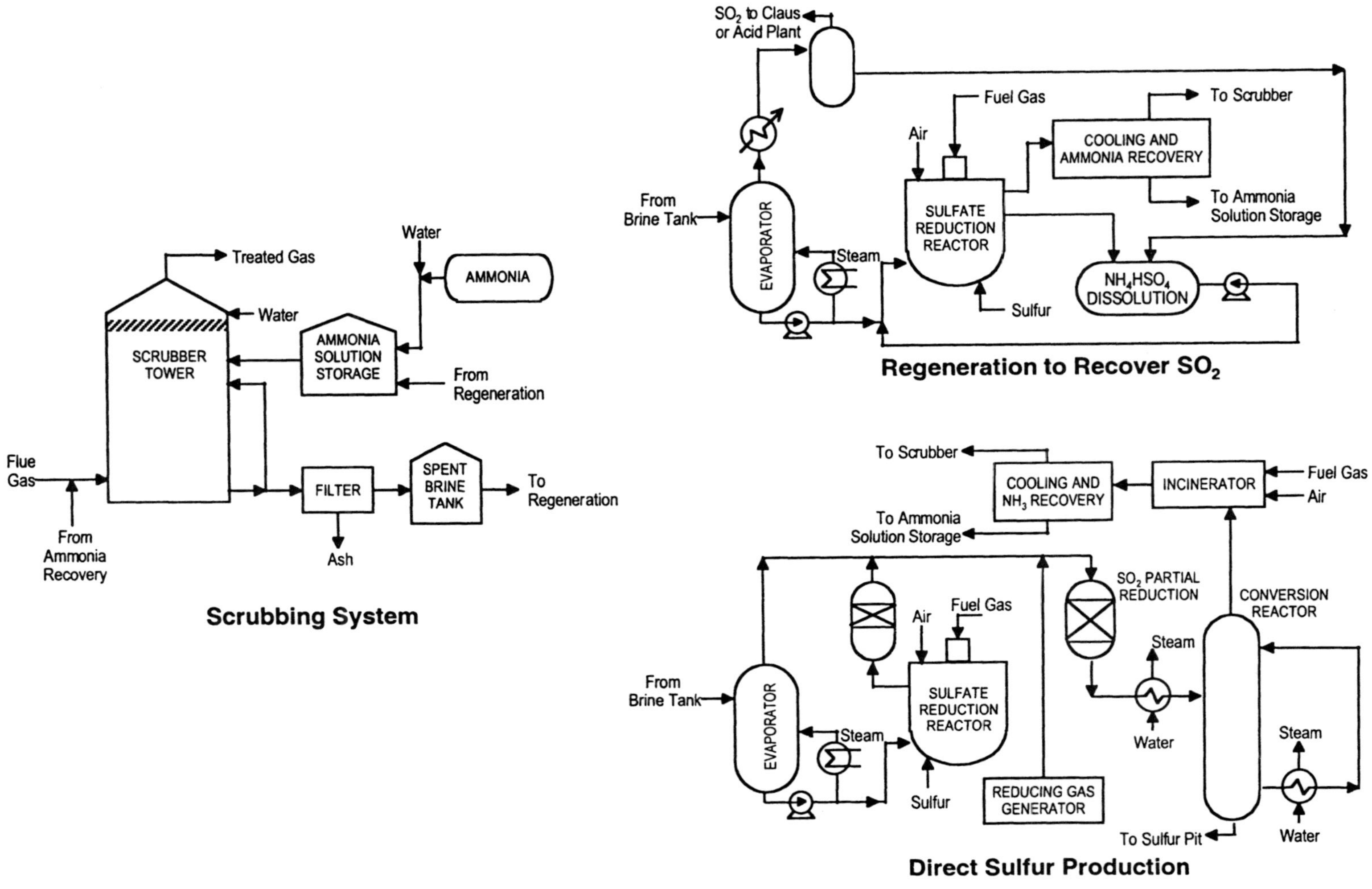

Regeneration to Recover SO$_2$

Direct Sulfur Production

Scrubbing System

Tung Process

Developmental
Absorption of SO_2 in sodium sulfite/sulfite solution from which the absorbed SO_2 is extracted by an organic solvent regenerating sulfite solution and producing a rich solvent liquor that is steam stripped to recover SO_2.

Process Characteristics
SO_2 Sorbent: Sodium sulfite/bisulfite
Principal Raw Materials: Caustic or soda ash makeup; Organic solvent makeup; (Lime for prescrubber blowdown treatment)
Potentially Saleable Byproducts: SO_2-rich gas for conversion to sulfuric acid or sulfur
Solid Wastes: Wastes from the prescrubber blowdown treatment
Liquid Effluents: Purge liquors to control soluble impurities buildup (e.g., sulfate and oxidized organic solvent)
Added Gaseous Emissions: Potential for solvent losses
Inlet SO_2 Levels: 1,000 to 3,600 ppm (Tested)
SO_2 Removal: 95-99% (Claimed)
NO_x Removal: up to 90% (Claimed)
Particulate Removal: Upstream decoupled particulate control is required

Commercial Application
Number/Types: None
Locations: NA
First Deployment: NA
Current Status: Inactive

Process Description
Flue gas is first passed through a prescrubber to remove particulate and soluble impurties which are discharged through a prescrubber blowdown treatment system. The humidified gas is then contacted in a high efficiency absorber (e.g., tray or packed tower) with a solution of sodium sulfite/bisulfite that absorbs SO_2. Spent absorbent liquor rich in bisulfite is fed to a multistage countercurrent extraction system. During extraction, SO_2 is transferred to an organic solvent, regenerating the aqueous sulfite solution which is then returned to the scrubber feed tank. The SO_2-rich solvent is steam stripped to evolve SO_2-rich gas from which water vapor is condensed prior to further processing to sulfuric acid or elemental sulfur. Lean solvent from the stripper bottoms is returned to the solvent storage tank for reuse. Purge streams are required to control buildup of both soluble and insoluble impurities (e.g., tramp dust/particulate; sulfate formed from sulfite oxidation; and oxidized solvent). In order to minimize these purge streams, most vessels are blanketed with nitrogen gas.

Advantages/Disadvantages

Advantages
- Clear solution scrubbing minimizes scaling and plugging problems
- High SO_2 removal efficiency potential over a range of inlet SO_2
- Produces byproduct SO_2
- Very low steam requirements relative to direct steam stripping systems such as the Wellman Lord Process

Disadvantages/Limitations
- Organic solvent extraction systems can be sensitive to impurities and "temperamental" to operate leading to higher than expected solvent losses/makeup requirements
- Many purge streams are required to control buildup of inerts and impurities (prescrubber blowdown, sulfate from SO_2 oxidation, other soluble impurities and oxidized solvent)
- Need for nitrogen blanketing

Principal Developers
Raycon Research and Development

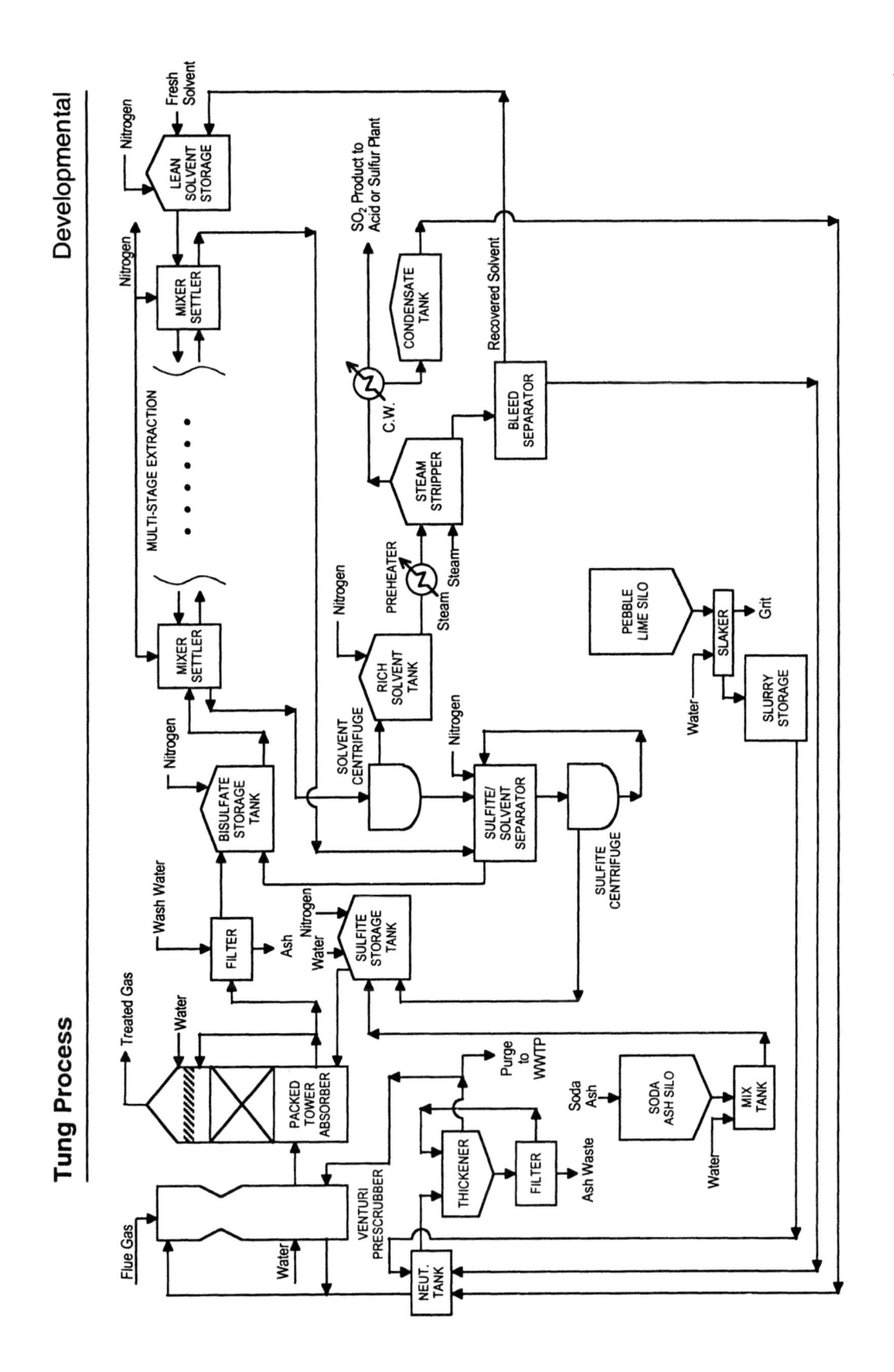

Byproduct Processes   121

Ionics Process

Developmental
Absorption of SO$_2$ in sodium sulfite/bisulfite solution which is then acidified with sulfuric acid releasing SO$_2$-rich gas for further processing and a sulfate solution that is regenerated electrolytically to sodium hydroxide.

Process Characteristics
SO$_2$ Sorbent: Sodium sulfite/bisulfite or sulfite/hydroxide
Principal Raw Materials: Caustic; Lime (for prescrubber blowdown treatment)
Potentially Saleable Byproducts: SO$_2$-rich gas for conversion to sulfuric acid or sulfur; 10% sulfuric acid; (Could require conversion gypsum)
Solid Wastes: Waste from prescrubber blowdown treatment
Liquid Effluents: Purge liquor for soluble impurities buildup
Added Gaseous Emissions: Hydrogen from electrolytic calls; Oxygen from electrolytic calls
Inlet SO$_2$ Levels: 1,000 to 3,600 ppm (Tested)
SO$_2$ Removal: up to 95% (Tested)
NO$_x$ Removal: Nil
Particulate Removal: Upstream decoupled particulate control required

Commercial Application
Number/Types: None
Locations: NA
First Deployment: NA
Current Status: Inactive (~2,000 cfm Pilot Plant Operated)

Process Description
Flue gas is contacted with a solution of sodium sulfite/bisulfite or sodium sulfite and hydroxide for SO$_2$ absorption. Spent absorbent liquor is continuously withdrawn from the scrubber and sent to a regeneration circuit where the solution is contacted with sulfuric acid releasing SO$_2$ and producing sodium sulfate. The resulting sodium sulfate solution is then passed to a set of electrolytic cells where the sodium sulfate is converted to caustic and sulfuric acid (along with hydrogen and oxygen). Two types of cells are used. One produces impure, dilute sulfuric acid that is recycled for acidification of the absorber bleed. A second, more efficient set of cells is used to more completely regenerate sodium sulfate to 10% sulfuric acid. The number of cells required and amount of 10% acid production represents the extent of oxidation and sulfur trioxide absorption. In the pilot plant, roughly one-third of the electrolyzer cells were of the more efficient type.

Advantages/Disadvantages

Advantages
- Clear solution scrubbing minimizes scaling and plugging problems
- High SO$_2$ removal efficiency potential over a range of inlet SO$_2$
- Produces byproduct SO$_2$

Disadvantages/Limitations
- Full scale operating parameters have not been demonstrated in integrated operations
- Production of purge liquor and solids
- 10% byproduct sulfuric acid may have limited market value and may require conversion to gypsum

Principal Developers
Stone and Webster and Ionics Inc.

Byproduct Processes

Ionics Process — Sodium Solution with Electrolytic Regeneration — Developmental

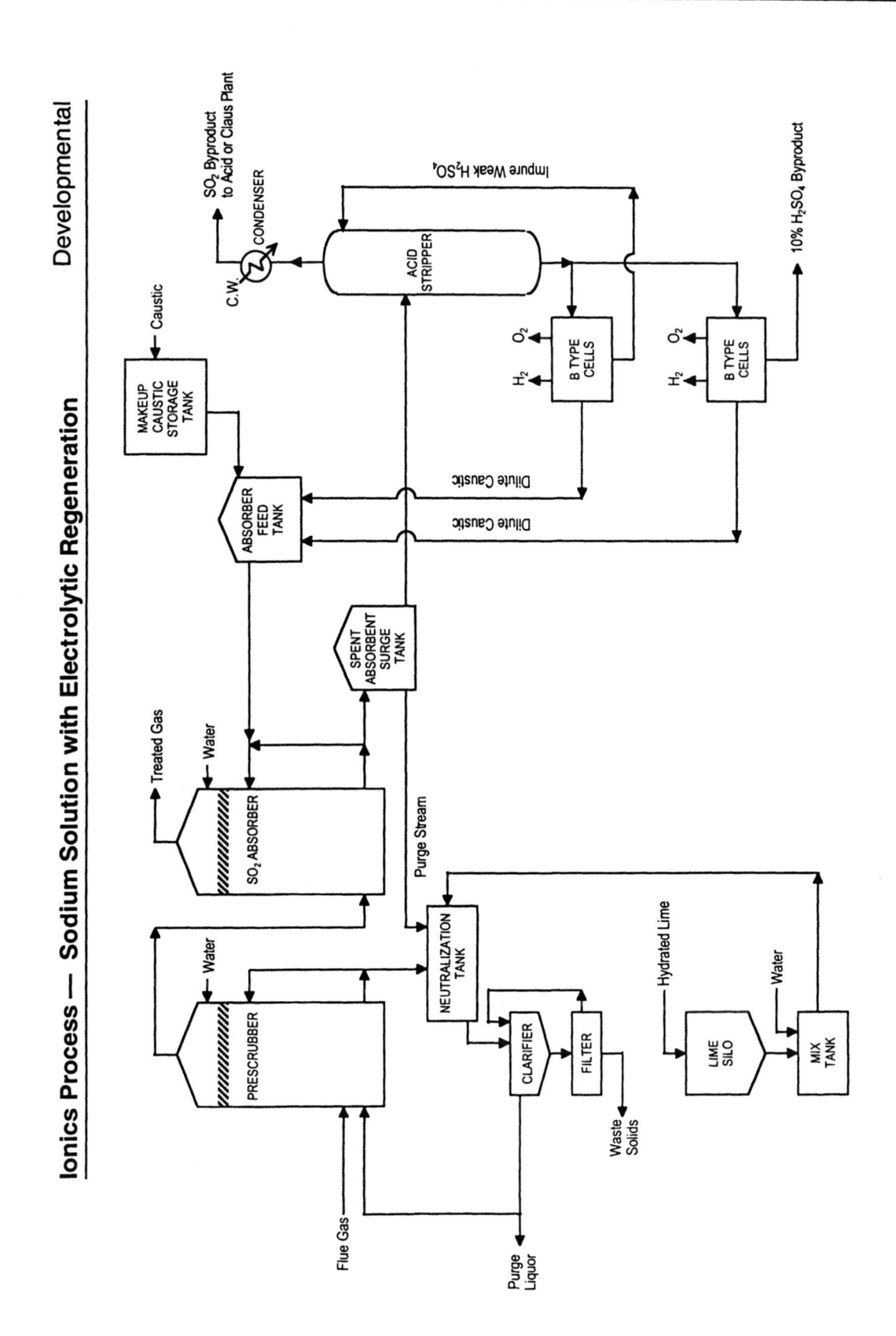

SOXAL Process

Developmental
Absorption of SO_2 in sodium sulfite solution which is then subjected to electrodialysis converting bisulfite to sulfurous acid from which SO_2-rich gas is stripped and regenerated sodium sulfite recycled to the scrubber.

Process Characteristics
SO_2 Sorbent: Sodium sulfite/bisulfite
Principal Raw Materials: Caustic; (Lime for prescrubber blowdown treatment); (Urea and methanol if NO_x control were effected - proposed)
Potentially Saleable Byproducts: SO_2-rich gas for conversion to sulfuric acid or sulfur; Sodium sulfate solids
Solid Wastes: Waste from prescrubber blowdown treatment
Liquid Effluents: Purge liquor for soluble impurities buildup
Added Gaseous Emissions: None
Inlet SO_2 Levels: up to 3,600 ppm (Tested)
SO_2 Removal: 99% (Claimed)
NO_x Removal: Potential for up to 90% (Claimed but not tested)
Particulate Removal: Upstream decoupled particulate control is required

Commercial Application
Number/Types: None
Locations: NA
First Deployment: NA
Current Status: Inactive (~10,000 cfm Pilot Plant Operated)

Process Description
Flue gas is contacted with a solution of sodium sulfite/bisulfite which absorbs SO_2. Spent absorbent liquor rich in bisulfite is continuously withdrawn to a regeneration circuit consisting of one or two stages. The first stage is a stack of electrodialysis cells and a steam stripper. The first stage electrodialysis cells convert bisulfite to sulfurous acid and sodium sulfite. Sodium sulfite is recycled to the scrubber circuit and the sulfurous acid is stripped with steam to produce water and SO_2-rich gas for further processing to sulfuric acid or sulfur. If the second stage is not used, sodium sulfate formed from the oxidation of absorbed SO_2 or absorption of SO_3 must be purged from the system, either as a liquor or after crystallization. The second stage consists of another stack of electrodialysis cells to convert sodium sulfate to dilute caustic, which is recycled to the scrubber, and dilute sulfuric acid. The acid would normally be reacted with limestone to produce gypsum. Other soluble impurities (e.g., chlorides, nitrates from combined NO_x control if implemented) need to be purged either with soluble sodium sulfate (one stage system) or in neutralized hydrochloric acid (two stage system). NO_x removal capability is also claimed, but has yet to be tested. NO_x control would involve economizer injection of urea (to reduce NO_x to N_2) followed by injection of methanol to oxidize residual NO to NO_2 (and also oxidize ammonia slip) which is then removed in the scrubber where it reacts to produce sodium nitrate, sulfate, N_2 and H_2.

Advantages/Disadvantages

Advantages
- Clear solution scrubbing minimizes scaling and plugging problems
- High SO_2 removal efficiency potential over a range of inlet SO_2
- Produces byproduct SO_2
- Potential for integrated SO_2 and NO_x control

Disadvantages/Limitations
- Stage two regeneration has not been tested
- Full scale operating parameters have not been demonstrated in integrated operations
- Production of purge liquor and prescrubber treatment solids

Principal Developers
AlliedSignal/Aquatech

Byproduct Processes 125

SOXAL Process — Sodium Solution with Electrodialysis Regeneration   Developmental

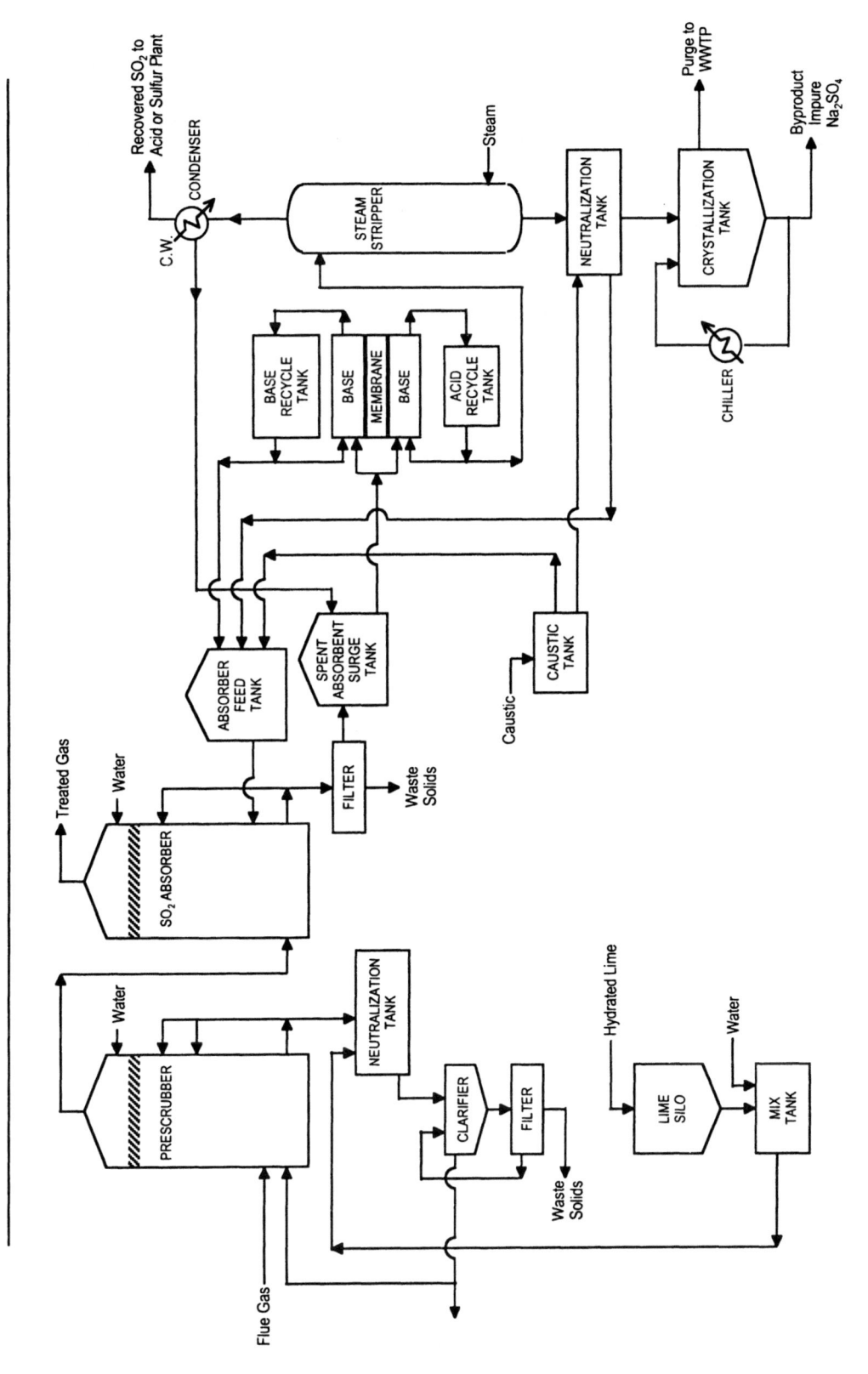

Zinc Oxide Process (Sulfite Solution Absorption)

Developmental
Absorption of SO_2 in a sodium solution, precipitation of the dissolved SO_2 from the spent solution using zinc oxide, then calcination of the zinc sulfite to evolve SO_2-rich gas for further processing to acid or sulfur.

Process Characteristics
SO_2 Sorbent: Sodium sulfite/bisulfite
Principal Raw Materials: Lime; Zinc Oxide; Soda ash
Potentially Saleable Byproducts: SO_2-rich gas for conversion to sulfuric acid or sulfur
Solid Wastes: Impure gypsum
Liquid Effluents: Purge liquor to control buildup of soluble impurities
Added Gaseous Emissions: None
Inlet SO_2 Levels: Not defined
SO_2 Removal: Not defined
NO_x Removal: Nil
Particulate Removal: Upstream decoupled particulate removal required

Commercial Application
Number/Types: None
Locations: NA
First Deployment: NA
Current Status: Inactive (Reached pilot plant stage in 1940s)

Process Description
Flue gas is first contacted with a solution of sodium sulfite/bisulfite that absorbs SO_2 increasing the bisulfite concentration. Spent absorbent liquor is first passed through a clarifier to remove any particulate matter picked up in the absorber. The clarified liquor is sent to a reactor where it is treated with ZnO. The zinc oxide precipitates zinc sulfite and regenerates sodium hydroxide which neutralizes the bisulfite to sulfite. The zinc sulfite is separated by thickening and filtration and the regenerated absorbent is returned to the absorber. The filter cake is dried and calcined releasing SO_2 and water vapor. Product SO_2 is cooled and dried. Recovered zinc oxide is recycled to the regeneration reactor. SO_2 that is oxidized in the process and absorbed SO_3 represent an "inert" burden which must be removed. This is accomplished by treating a slipstream from the clarifier with lime which precipitates calcium sulfite. The thin calcium sulfite slurry is then thickened in the clarifier and the thickened slurry is acidified with a portion of the product SO_2. This redissolves the calcium and precipitates calcium sulfate, which is then removed with any ash by thickening and filtration. Resulting "desulfated" solution is recycled to the process through the lime reactor.

Advantages/Disadvantages

Advantages
- Clear solution scrubbing minimizes scaling and plugging problems
- High SO_2 removal efficiency potential over a range of inlet SO_2
- Produces byproduct SO_2

Disadvantages/Limitations
- Recent economic evaluations find the process capital costs to be high relative to other byproduct recovery processes
- Full scale operating parameters have not been demonstrated in integrated operations

Principal Developers
University of Illinois

Byproduct Processes

Zinc Oxide Process (Sulfite Solution Absorption)   Developmental

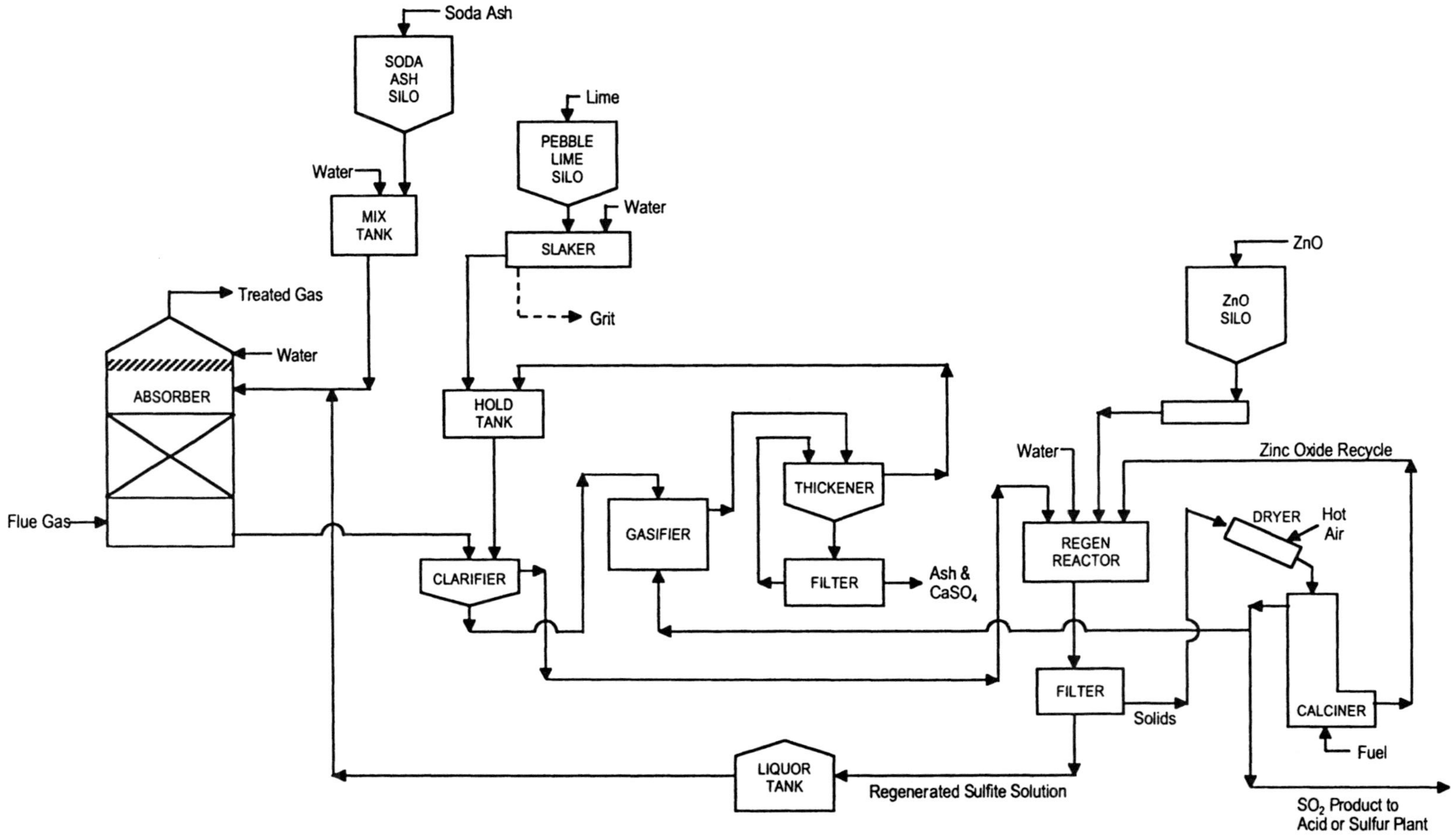

Sorbtech Process

Developmental
Sorption of SO_2, NO_X and some SO_3 and HCl in a magnesium oxide/silicate sorbent that is regenerated by heating under either reducing (for NO_x reduction) or oxidizing conditions releasing SO_2 for further processing.

Process Characteristics
SO_2 Sorbent: Mag*Sorbent® (Combined MgO and vermiculite)
Principal Raw Materials: Mag*Sorbent®; Natural gas
Potentially Saleable Byproducts: SO_2-rich gas for conversion to acid or sulfur (Oxidizing regeneration); Combined SO_2, H_2S, and sulfur for sulfur production (Reducing regeneration); Spent sorbent as a fertilizer additive or soil amendment
Solid Wastes: None
Liquid Effluents: None
Added Gaseous Emissions: None
Inlet SO_2 Levels: ~2,500 ppm (Tested)
SO_2 Removal: 90+% (Tested)
NO_x Removal: 40-80% (Tested)
Particulate Removal: Upstream decoupled particulate control required

Commercial Application
Number/Types: None
Locations: NA
First Deployment: NA
Current Status: Active (Preliminary demonstration testing completed)

Process Description
After partial humidification, the flue gas is contacted in a vertical, cylindrical vessel with a specially designed radial-panel bed of slowly moving magnesium-based sorbent. The sorbent removes SO_2, much of the NO_x and other acid gases such as HCl. The sorbent is a patented mix of specially prepared magnesium oxide and vermiculite. The removal mechanism is a combination of chemical absorption/reaction and capillary condensation. Spent sorbent is continuously removed from the absorber and regenerated by heating with natural gas combusted with air. There are two approaches developed for regeneration. First is heating under reducing conditions at 750°C. This decomposes the magnesium sulfite and sulfate, releasing a combined gas of SO_2, H_2S and sulfur vapor, and reducing NO_x to nitrogen and water. The regeneration offgas must be processed in a modified Claus plant to produce elemental sulfur. Second is heating in an oxidizing atmosphere at 550°C. This releases SO_2, but does not completely decompose magnesium sulfate which then must be purged from the system at the rate at which oxidation and SO_3 absorption occur. NO_x is also not fully reduced, limiting the extent of overall NO_x removal performance. In either case, the regeneration circuit consists of a regenerator and a screening station where 5-20% of the sorbent is removed depending upon the type or regeneration. Fresh sorbent is then added prior to recycle to the absorber circuit.

Advantages/Disadvantages

Advantages
- Produces SO_2 byproduct gas
- Simultaneous SO_2 and NO_x removal
- Spent sorbent has potential byproduct value as fertilizer supplement or soil amendment
- Stack gas requires no reheat

Disadvantages/Limitations
- Potential for high sorbent losses, but relatively low cost sorbent

Principal Developers
SorbTech

Byproduct Processes

Sorbtech Process — Developmental

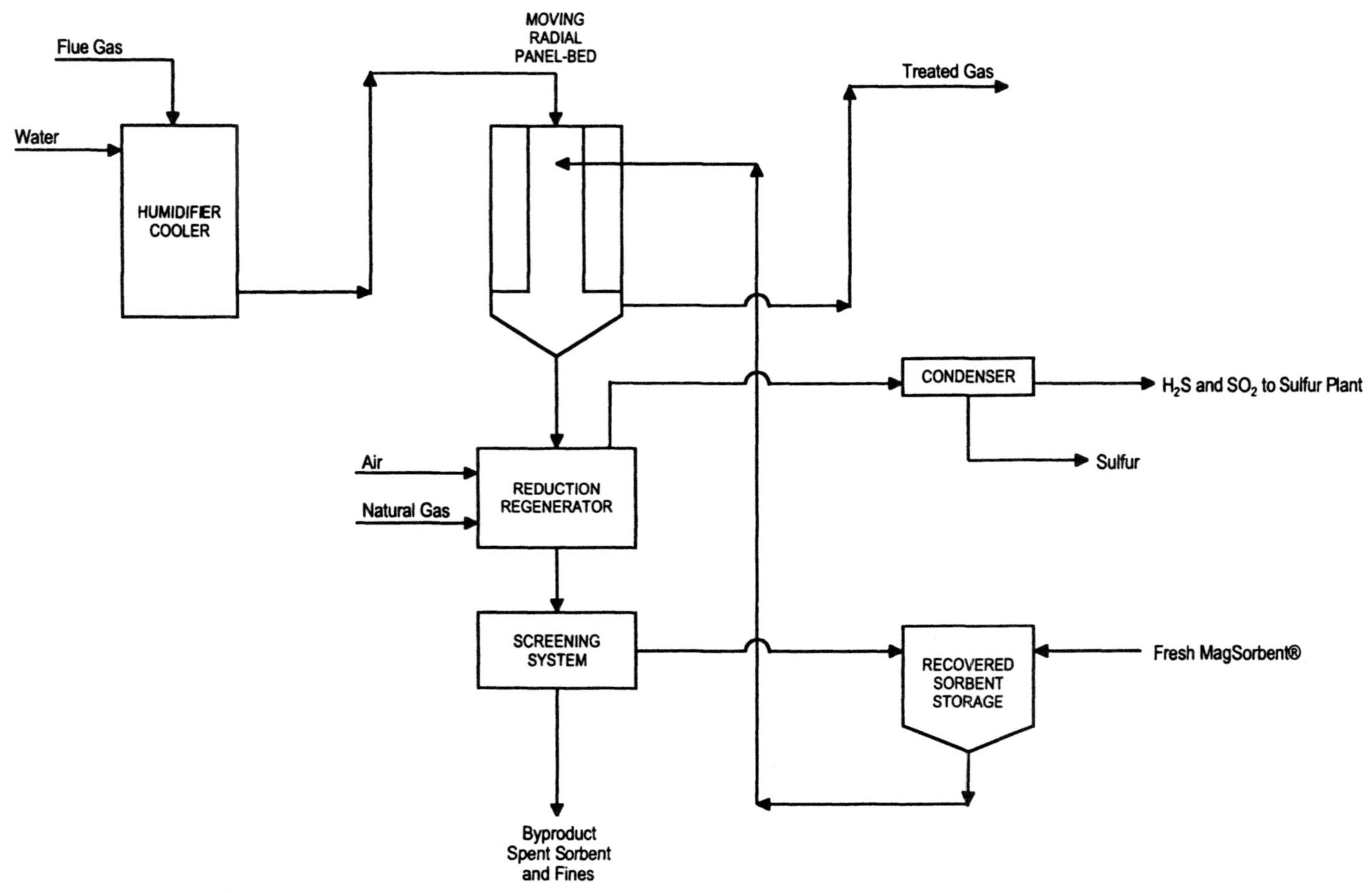

Inputs/outputs shown in the flow diagram:
- Flue Gas, Water → HUMIDIFIER COOLER
- → MOVING RADIAL PANEL-BED → Treated Gas
- → REDUCTION REGENERATOR (Air, Natural Gas) → SCREENING SYSTEM → Byproduct Spent Sorbent and Fines
- → CONDENSER → Sulfur; H$_2$S and SO$_2$ to Sulfur Plant
- Fresh MagSorbent® → RECOVERED SORBENT STORAGE

Citrate Process

Developmental
Absorption of SO$_2$ a solution of sodium citrate followed by reaction with hydrogen sulfide in an Aqueous Claus reactor to precipitate elemental sulfur and regenerate the absorbent solution.

Process Characteristics
SO$_2$ Sorbent: Sodium citrate solution
Principal Raw Materials: Caustic or soda ash; Citric acid; Reducing gas; Lime (for prescrubber blow down neutralization)
Potentially Saleable Byproducts: Elemental sulfur; Sodium sulfate
Solid Wastes: Wastes from the prescrubber blowdown treatment
Liquid Effluents: Purge liquors required to control soluble impurities buildup (e.g., chlorides)
Added Gaseous Emissions: Incinerator for sulfide offgases from sulfur precipitation reactors (could be discharged to the prescrubber)
Inlet SO$_2$ Levels: 2,000 to 30,000 ppm (Tested)
SO$_2$ Removal: 90+% (Tested)
NO$_x$ Removal: Nil
Particulate Removal: Upstream decoupled particulate control is required

Commercial Application
Number/Types: None
Locations: NA
First Deployment: NA
Current Status: Inactive (Demonstration completed on a smelter and ~50 MWe power plant)

Process Description
Flue gas is first passed through a prescrubber to remove particulate and soluble impurities that are discharged through a prescrubber blowdown treatment system. The humidified gas is then contacted in a high efficiency absorber (e.g., packed tower) with a solution of sodium citrate that absorbs SO$_2$. Spent absorbent liquor is fed to a closed, stirred vessel in which it is reacted with hydrogen sulfide to precipitate elemental sulfur. The reactor product slurry is then thickened and centrifuged to remove the sulfur from the regenerated citrate solution. A small portion of the regenerated solution is purged to control buildup of soluble impurities and the remainder of the recovered solution is returned to the absorber. In the final purification step, the sulfur is liquefied in an autoclave and the molten sulfur is separated from the residual citrate solution that is also returned to the process. Two-thirds of the molten sulfur product is then converted to H$_2$S for use in absorbent regeneration.

Advantages/Disadvantages

Advantages
- Clear solution scrubbing should minimize scaling in the absorber although plugging problems were encountered (see Limitations)
- Applicable to a wide range of inlet SO$_2$
- Produces byproduct sulfur directly

Disadvantages/Limitations
- Highly corrosive conditions in regeneration and recovery circuits with very expensive materials
- Plugging problems plagued demonstration operations due to inability to completely separate sulfur from the citrate solution
- Requires a purge stream to control impurities
- Highly complex process configuration
- Requires reducing gas

Principal Developers
Pfizer
U.S. Bureau of Mines

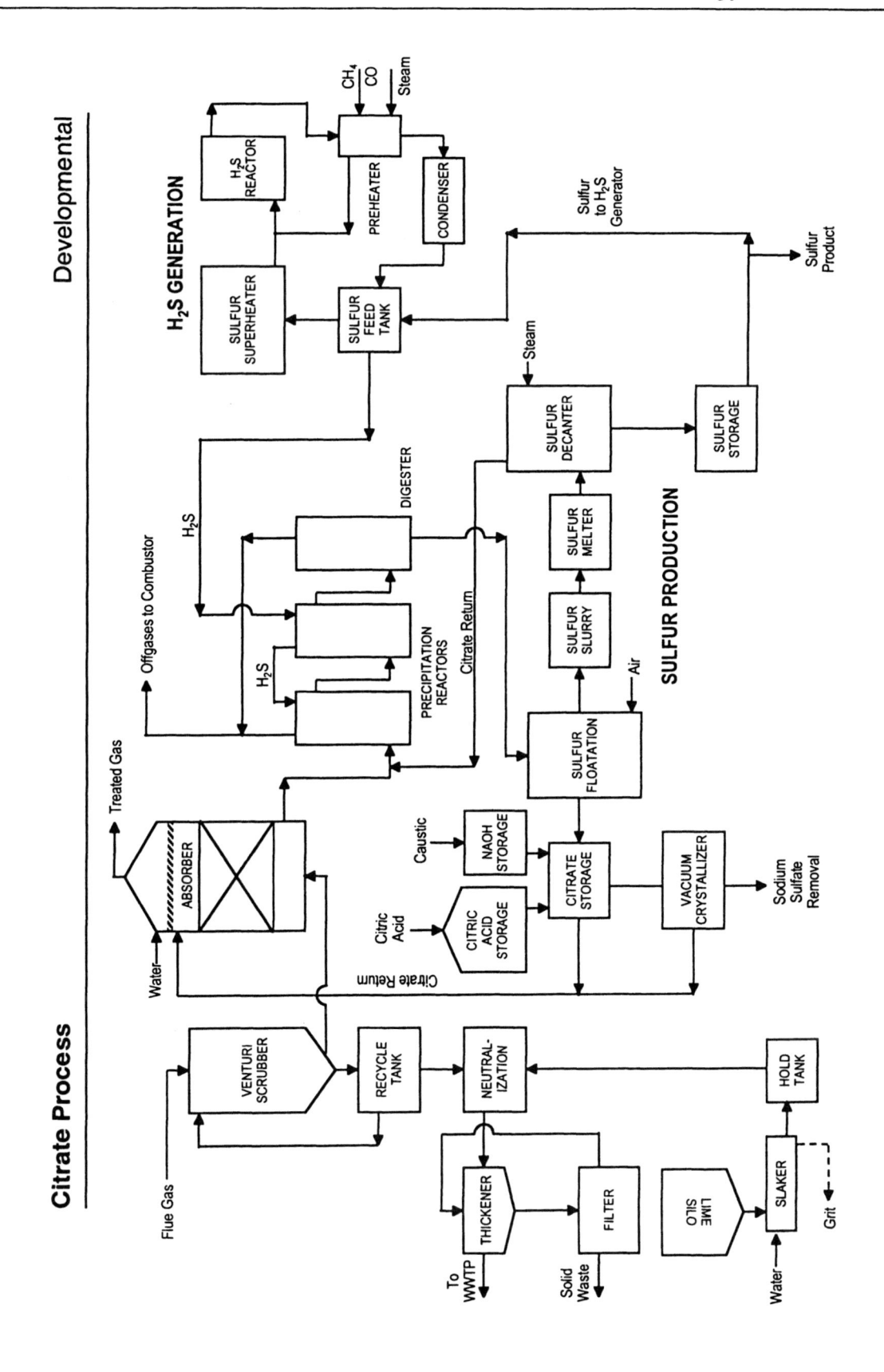

Byproduct Processes 131

Sulf-X Process

Developmental
Absorption of SO_2 (and NO_x) in an iron sulfide slurry that is then treated with sodium sulfide, the resultant solids dried and calcined with coke to regenerate the solution and produce sulfur.

Process Characteristics
SO_2 Sorbent: Aqueous slurry of FeS and $Fe(OH)_2$
Principal Raw Materials: Pyrites; Coke; Sodium sulfide; Lime (for prescrubber blowdown treatment)
Potentially Saleable Byproducts: Elemental sulfur
Solid Wastes: Wastes from the prescrubber blowdown treatment
Liquid Effluents: Purge liquors to control soluble impurities buildup (e.g., chlorides)
Added Gaseous Emissions: None
Inlet SO_2 Levels: 500 to ~3,000 ppm (Tested)
SO_2 Removal: 85-99% (Tested)
NO_x Removal: 65-90+%
Particulate Removal: Upstream particulate control is required

Commercial Application
Number/Types: None
Locations: NA
First Deployment: NA
Current Status: Inactive (Pilot plant operations completed)

Process Description
Flue gas is first passed through a prescrubber to remove particulate and soluble impurties which are discharged through a prescrubber blowdown treatment system. The humidified gas is then contacted with an aqueous slurry of ferrous sulfide and hydroxide. Ferrous sulfide is the primary absorbent; ferrous hydroxide serves to stabilize pH. Absorption produces a ferrous sulfate and a complex solution of iron sulfides. Spent solution is first treated with sodium sulfide to convert ferrous sulfate to sodium sulfate and iron sulfide. The solution is then dewatered and the solids dried and calcined at 1400°F in the presence of coke. Thermal treatment decomposes complex Fe_xS_y to simple iron sulfide and sulfur; and, the reducing conditions convert sodium sulfate to sodium sulfide. Elemental sulfur leaves the regenerator as a vapor and is condensed as the principal byproduct. Regenerated FeS and Na_2S are returned to the process. The process chemistry has the inherent capability for integrated, simultaneous removal of NO_x and it has been tested in both modes – FGD only and combined FGD and NO_x control. It was demonstrated to have the capability to convert over 90% of NO_x to elemental nitrogen. The major drawback, though, is the slow reaction rates that necessitate very large absorbers, probably making combined operation uneconomical on a commercial scale.

Advantages/Disadvantages

Advantages
- Direct production of sulfur
- Low steam requirements relative to other recovery technologies
- Absence of requirement for a reducing gas
- Potential for simultaneous, integrated SO_2 and NO_x control

Disadvantages/Limitations
- Slurry scrubbing which complicates absorption for high efficiency performance
- Operation of a high temperature calciner
- Handling of three different solids reagents
- Need for to circulate a complex slurry throughout much of the process
- A fairly complex processing scheme

Principal Developers
Pittsburgh Environmental and Energy Systems

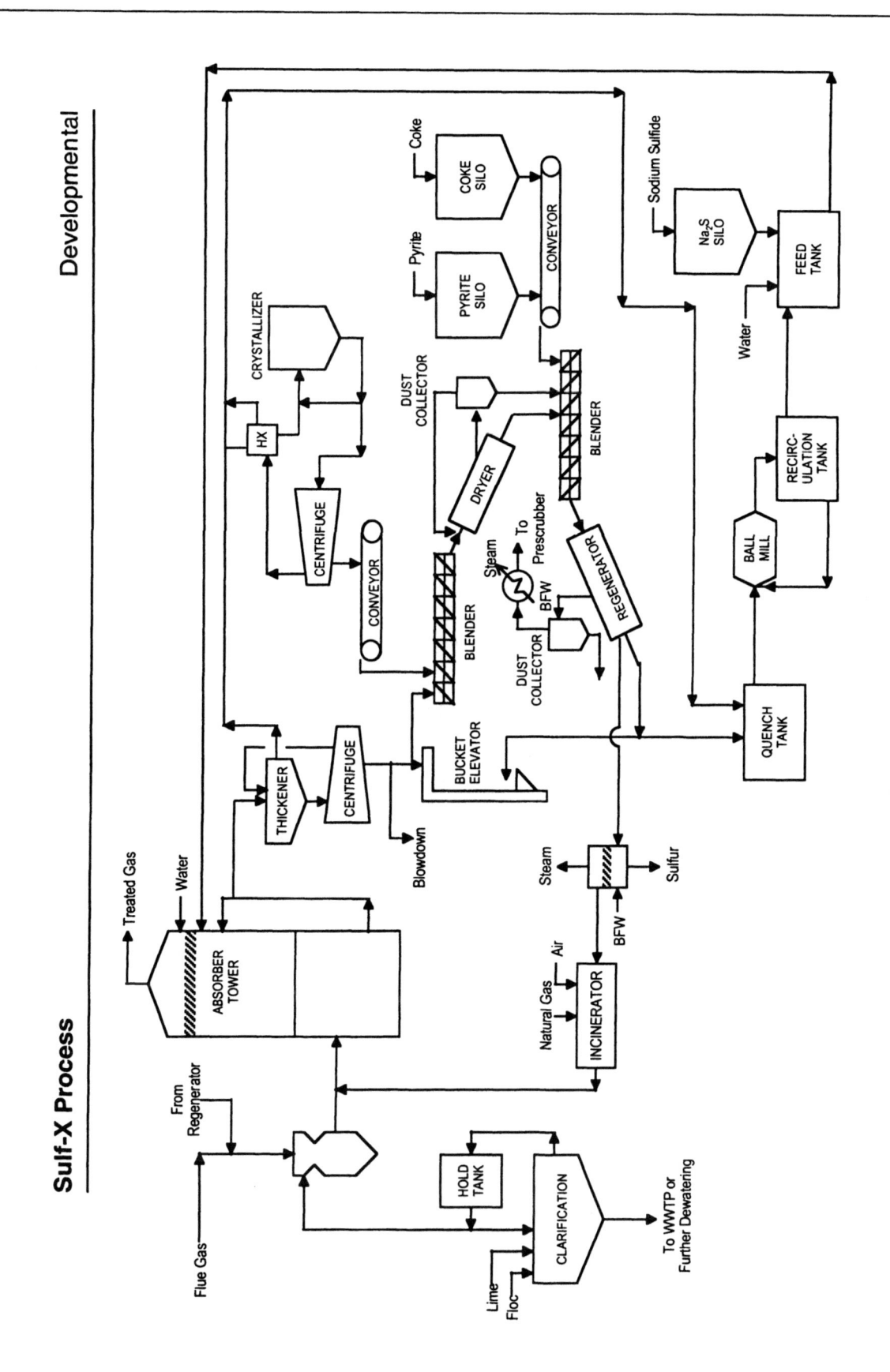

Byproduct Processes

Sulf-X Process — Developmental

Direct Gas Phase Reduction

Developmental
Simultaneous reduction of SO_x and NO_x in the flue gas to hydrogen sulfide and nitrogen, respectively, recovery of the hydrogen sulfide and production of elemental sulfur from the hydrogen sulfide-rich gas.

Process Characteristics
SO_2 Sorbent: Hydrogen and carbon monoxide
Principal Raw Materials: Steam; Natural Gas
Potentially Saleable Byproducts: Sulfur
Solid Wastes: None
Liquid Effluents: Sour water stripper blowdown
Added Gaseous Emissions: None
Inlet SO_2 Levels: ~3,000 ppm (Tested)
SO_2 Removal: 90+% Capability (Tested)
NO_x Removal: 85+% (Tested)
Particulate Removal: None

Commercial Application
Number/Types: None
Locations: NA
First Deployment: NA
Current Status: Inactive (Preliminary pilot plant testing completed)

Process Description
Of the processes researched, the Parsons Direct Reduction Process is the most recent and most thoroughly developed. In it, hot flue gas at ~750°F (e.g., upstream of a boiler air preheater) is contacted in a hydrogenation reactor where SO_x (both SO_2 and SO_3) and NO_x are simultaneously catalytically reduced to hydrogen sulfide and elemental nitrogen, respectively. The reducing gas for the hydrogenation reactor is syngas (hydrogen and carbon monoxide) produced from natural gas and steam. The gas is then cooled (in an air preheater and desuperheater coupled with a sour water stripper in the case of a boiler application) followed by conventional high efficiency particulate control, if required. Hydrogen sulfide is recovered from the cooled gas by any of a number of traditional H_2S absorption technologies, and the H_2S-rich gas is sent to a Claus plant for production of elemental sulfur. Claus plant tail gas is recycled to the hydrogenation reactor.

Advantages/Disadvantages

Advantages
- No handling of slurries
- Produces sulfur byproduct
- No solid waste products
- Simultaneous SO_2, SO_3, and NO_x control
- Stack gas requires no reheat

Disadvantages/Limitations
- High capital investment due mainly to the large number of successive unit operations for treating or conditioning the entire flue gas flow
- Purge liquors, particularly from the sour water stripper

Principal Developers
Chevron   Ontario Research Foundation   Parsons

Byproduct Processes

Parsons Process — Direct Gas Phase Reduction

Developmental

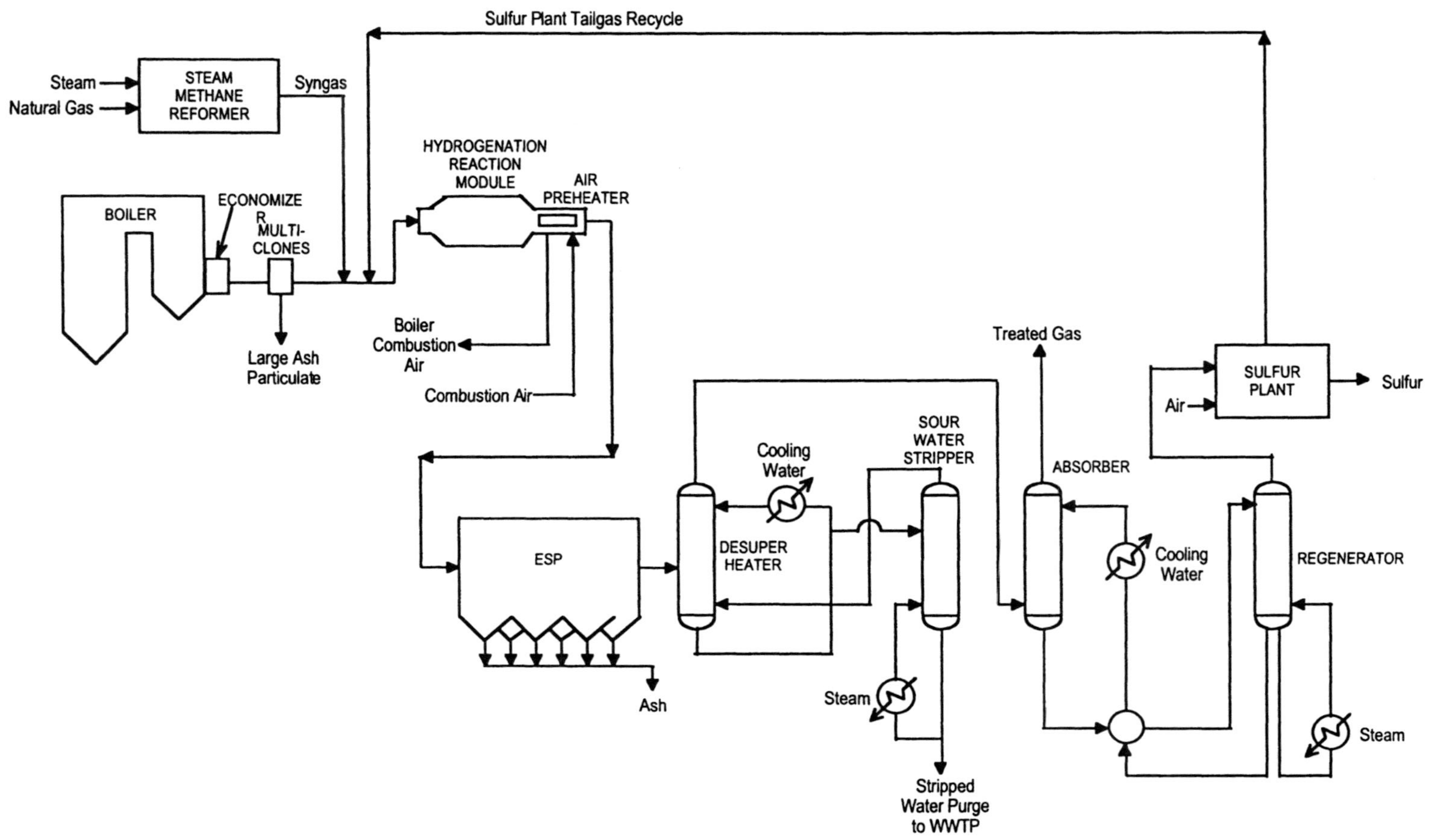

Copper Oxide Recovery Process

Developmental
Flue gas with ammonia added is contacted with copper oxide sorbent that absorbs SO_2 and acts as a catalyst for converting NO_x to nitrogen. Sorbent is regenerated with reducing gas to produce SO_2-rich byproduct gas.

Process Characteristics
SO_2 Sorbent: Copper oxide supported on alumina
Principal Raw Materials: Copper oxide sorbent; Ammonia; Steam; Natural gas
Potentially Saleable Byproducts: SO_2-rich gas for conversion to acid or sulfur
Solid Wastes: None
Liquid Effluents: None
Added Gaseous Emissions: Reduction reactor offgas incinerator (could be sent to the sorption unit)
Inlet SO_2 Levels: 1,000 to 3,500 ppm (Tested)
SO_2 Removal: 90-99% (Tested)
NO_x Removal: 90-95% (Tested)
Particulate Removal: Upstream decoupled, dry particulate control generally required

Commercial Application
Number/Types: None
Locations: NA
First Deployment: NA
Current Status: Active

Process Description
Copper oxide-based processes have been under development since the early 1970s and have yet to achieve fulfillment of early promises. As now configured, it is utilizes a gamma alumina substrate impregnated with copper oxide as a sorbent for SO_2 and a catalyst for NO_x. The absorber operates at temperatures near 750°F. Removal efficiencies depend upon reactor design and form of the sorbent/catalyst. Reactor designs now include moving beds, fluid beds and gas suspension (entrained bed) contactors. The moving bed and fluid bed units use pelletized sorbent whereas the entrained bed approach uses powdered sorbent. Copper oxide absorbs SO_2 producing calcium sulfate. Both copper oxide and copper sulfate act as catalysts for conversion of NO_x to nitrogen. Spent sorbent is passed to a regenerator where reducing gas consisting of hydrogen, carbon monoxide, and/or methane is used regenerate the sorbent. The temperature of regeneration depends on the type of reducing gas (e.g., the regenerator is operated slightly hotter when methane is used). During regeneration, the copper sulfate is converted to elemental copper and SO_2. Concentrated SO_2 is then further processed for conversion to sulfur or sulfuric acid. Regenerated sorbent is returned to the contactor where the regenerated copper oxide is immediately converted to copper oxide and reactivated by oxygen in the flue gas.

Advantages/Disadvantages

Advantages
- Produces SO_2-rich byproduct gas
- No significant solid wastes or liquid effluents
- No slurries or solutions to deal with
- Stack gas requires no reheat

Disadvantages/Limitations
- Complicated process configuration involving movement of solids and combustible gases
- Potential for high sorbent attrition losses
- Need to handle ammonia, a highly regulated hazardous chemical

Principal Developers
Air Products (Discontinued)
Exxon (Discontinued)

Shell (Inactive)
UOP (Inactive)

U.S. DOE/PETC with Thermo Electron (Active)

Byproduct Processes

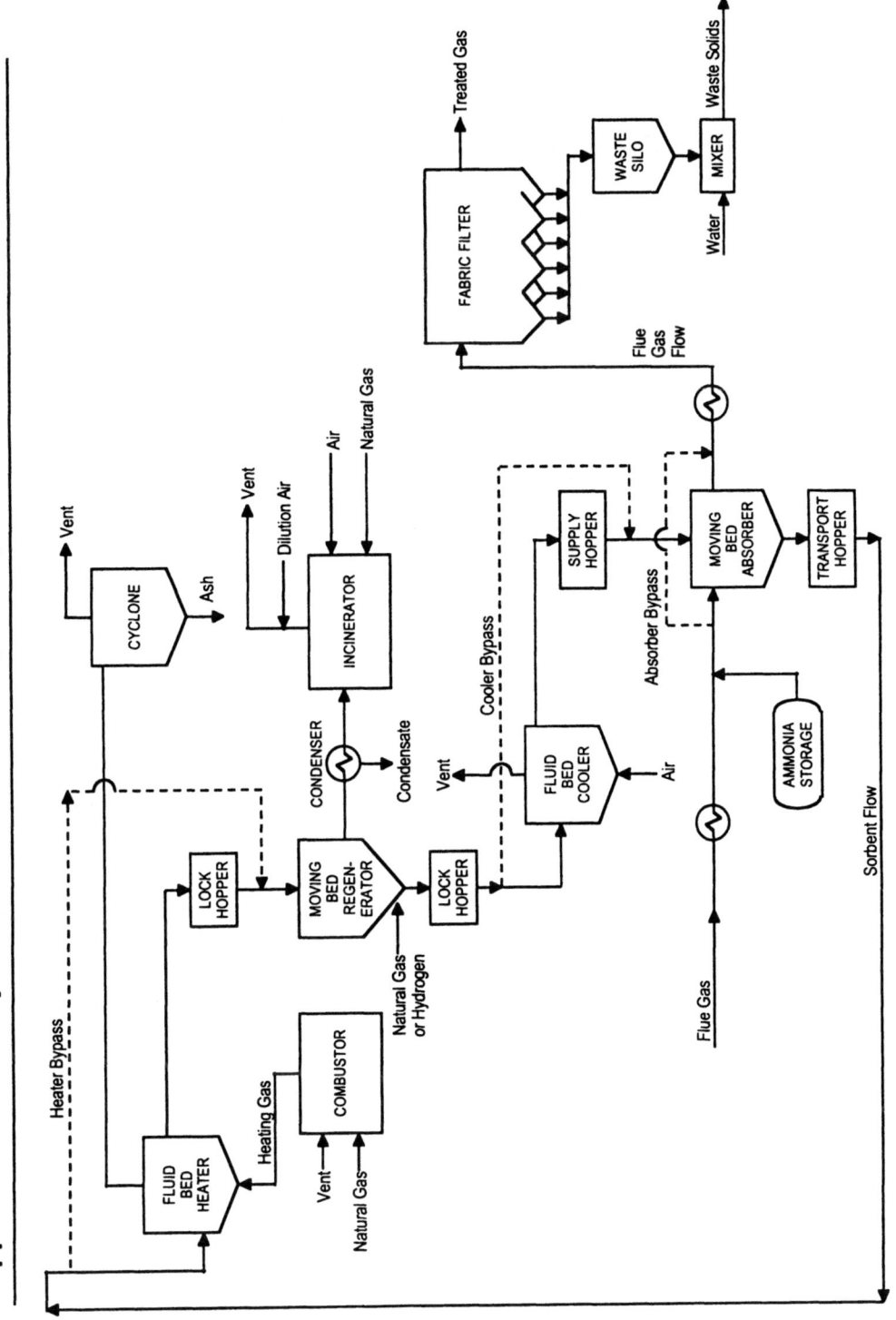

Copper Oxide Recovery Process—PETC Moving Bed System

Developmental

NOXSO Process

Developmental
Simultaneous removal of SO_x and NO_x in a gas/solids contactor by sorption on alkalized alumina that is regenerated with reducing gas to produce a mixed H_2S/SO_2 gas for conversion to sulfur in a Claus plant.

Process Characteristics
SO_2 Sorbent: Alkalized alumina solid
Principal Raw Materials: Steam; Natural gas; Sorbent makeup (alkalized alumina)
Potentially Saleable Byproducts: Sulfur
Solid Wastes: Spent sorbent – may be marketable
Liquid Effluents: None
Added Gaseous Emissions: Auxiliary natural gas heater offgas
Inlet SO_2 Levels: ~2,500 ppm (Tested)
SO_2 Removal: 90-95+% Capability (Tested)
NO_x Removal: 70-85+% (Tested)
Particulate Removal: None

Commercial Application
Number/Types: None
Locations: NA
First Deployment: NA
Current Status: Inactive (Demonstration testing completed; NOXSO suspended operation)

Process Description
Development of this technology predates the 1970s with early work by the U.S. Bureau of Mines which has subsequently been discontinued. A later version under development by the NOXSO Corporation began in the late 1970s followed by more recent efforts in a joint venture with FLS Miljo based upon modifications primarily geared toward the European market. The NOXSO process uses pellets of alumina layered with sodium aluminate, the active sorbent. Flue gas at typical 250-300°F is contacted in a moving or fluid bed bed with the pelletized sorbent which removes SO_2 and NO_x, forming sodium sulfate, sulfite and nitrate. Spent sorbent separated from the absorption reactor is regenerated in several steps. First, the sorbent is heated with air to about 1,150°F which strips NO_x and a small amount of SO_2. Hot air is supplied by regenerative heating-cooling system augmented with a natural gas fired heater. The stripped gas is sent to a combustion burner where staged burning reduces the NO_x to nitrogen. The sorbent is then sent to a multistage regenerator where it is reacted with steam and natural gas which produces H_2S and SO_2 that is sent to a Claus plant for conversion to elemental sulfur. The regenerated sorbent is recycled to the absorber. The principal difference between the traditional NOXSO process and the NOXSO/FLS Miljo modification is the use of powdered sorbent in the FLS Miljo gas suspension absorber rather than a pelletized sorbent in the NOXSO fluid or moving bed.

Advantages/Disadvantages

Advantages
- Produces sulfur byproduct
- No solid wastes except expended sorbent bleed
- Simultaneous SO_2 and NO_x control
- Stack gas requires no reheat

Disadvantages/Limitations
- Relatively high capital investment
- Purge liquors, particularly from the sour water stripper
- Generally high attrition rates of sorbent material equivalent to one complete turnover per year

Principal Developers

| NOXSO Corp. | NOXSO/FLS Miljo | U.S. Bureau of Mines |
| (Moving bed process) | (Gas Suspension Absorber) | (Development discontinued) |

Byproduct Processes

NOXSO/FLS-MILJO Process — Developmental

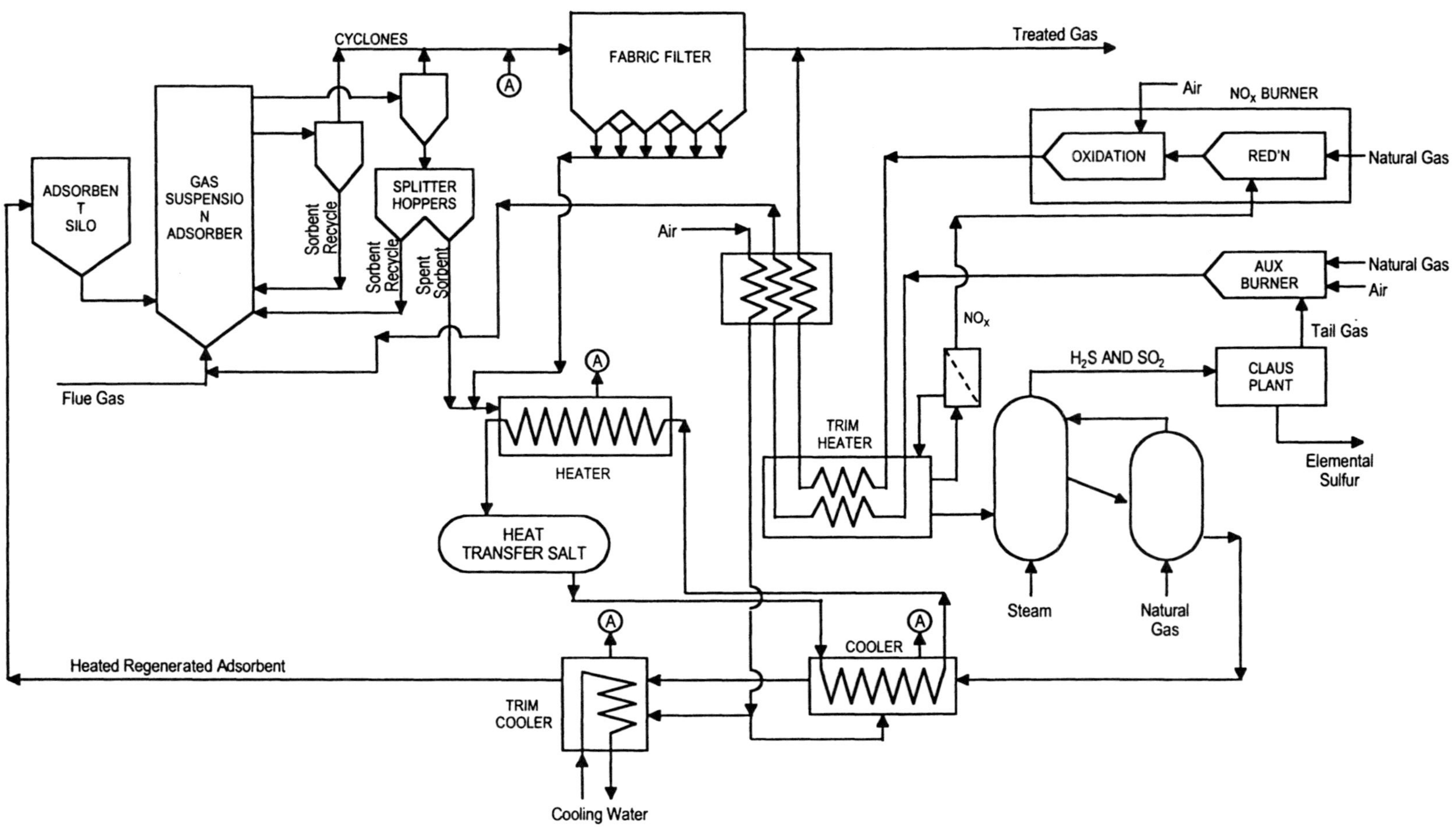